MW00620645

Follow the Science

ALSO BY SHARYL ATTKISSON

Slanted: How the News Media Taught Us to Love Censorship and Hate Journalism

The Smear: How Shady Political Operatives and Fake News Control What You See, What You Think, and How You Vote

Stonewalled: My Fight for Truth Against the Forces of Obstruction, Intimidation, and Harassment in Obama's Washington

Follow the Science

How Big Pharma Misleads, Obscures, and Prevails

Sharyl Attkisson

HARPER

An Imprint of Harper Collins*Publishers*

AUTHOR'S NOTE

The content of this book is based on my own opinions, experiences, and observations.

HarperCollins books may be purchased for educational, business, or sales promotional use. For information, please email the Special Markets Department at SPsales@harpercollins.com.

FIRST EDITION

Library of Congress Cataloging-in-Publication Data
Names: Attkisson, Sharyl, 1961– author.
Title: Follow the science : how big pharma misleads, obscures, and prevails / Sharyl Attkisson.
Identifiers: LCCN 2024011156 (print) | LCCN 2024011157 (ebook) | ISBN 9780063314917 (hardcover) | ISBN 9780063314924 (ebook)
Subjects: LCSH: Pharmaceutical industry—Corrupt practices.
Classification: LCC HD9665.5 .A86 2024 (print) | LCC HD9665.5 (ebook) | DDC 338.4/76151—dc23/eng/20240524
LC record available at https://lccn.loc.gov/2024011156
LC ebook record available at https://lccn.loc.gov/2024011157

24 25 26 27 28 LBC 5 4 3 2 1

With gratitude to family and friends who made this book possible. Thanks to the many brave, informative sources who provided insight for my reporting. Some proceeds from sales of this book support the professional Sharyl Attkisson ION Awards for Investigative and Original News reporting; the student ION Awards at the University of Florida Journalism and Communications College, and Diablo Valley College; the Brechner Center for Freedom of Information; and other independent reporting causes.

Remembering, with honor, Army Reservist Rachael Lacy. The military initially covered up the cause of her 2003 death. Investigating her case taught me to rightly question scientific matters I'd previously accepted unskeptically.

No legitimate scientist ever declared science to be "settled."

Contents

Introduction

We're deeply mired in a serious crisis that is touching all of us and growing worse. It is virtually ignored by our elected officials and popular media—but for their efforts to bury it.

It's a health crisis that encompasses an epidemic of chronic and acute diseases, an explosion in disorders related to Covid and Covid vaccines, historic levels of mental illness, and more young people than ever afflicted by brand-new diseases or those previously unheard-of in children.

We could fix these emergencies, but they are nowhere to be found on the national agenda set by our health agencies and major political figures. They're too busy spending your tax dollars on their own priorities.

Their priorities include deflecting from the crisis by feeding us a steady stream of propaganda and misinformation. Creating a market for, selling, and defending medicine. And smearing anyone who stands in the way. Why they do this, contrary to all of our best interests, is at the heart of this book.

Death by Prescription Drugs

Long before the opioid crisis, prescription drugs were a top cause of death. Today, they are number one, estimates Danish physician Dr. Peter C. Gøtzsche. In 2014, Dr. Gøtzsche helped start an independent collaborative, hoping to cut through the government-medical industry's hold on our health information. At the time, he wrote:

> Our drug agencies . . . rely on fake fixes . . . a long list of warnings, precautions, and contraindications for each drug, although they know that no doctor can possibly master all of these. Major reasons for the many drug deaths are impotent drug regulation, widespread crime that includes corruption of the scientific evidence about drugs and bribery of doctors, and lies in drug marketing, which is as harmful as tobacco marketing and, therefore, should be banned. We should take far fewer drugs, and patients should carefully study the package inserts of the drugs their doctors prescribe for them and independent information sources about drugs such as Cochrane reviews [referring to a scientific collaborative group], which will make it easier for them to say "no thanks."

Ten years later, drug ads are more ubiquitous than ever, and fatalities from prescription medicine are the top cause of death, Dr. Gøtzsche says, often written off to other factors. For example, thousands die of peptic ulcer complications caused by taking common non-steroidal anti-inflammatory drugs, or NSAIDs. Others die after falls prompted by medicine they were prescribed.

But because the solutions don't rest with consuming more drugs that deposit profits into pharmaceutical industry coffers, death-by-prescription-drug doesn't draw near the attention that other maladies do—which is par for the course.

It's not as if we haven't tried to get a grasp on our own health. But for every safety measure enacted on our behalf, there are well-funded and powerful interests working to undermine it. Warning labels are required for drugs, including vaccines. But doctors who prescribe them usually don't tell you what's on the labels, encourage you to read them, or later find out if you suffered any ill effects. And if you end up back in the clinic or in the ER with an adverse event, odds are the doctors won't recognize or acknowledge it as drug-related. They certainly don't report it to the federal databases designed to track and unearth new side effects. Why?

Speaking of those federal databases, when they do reveal patterns of injuries pointing to possible drug dangers, we're left to wonder why the data is collected at all. That's because government and pharmaceutical interests work in an organized fashion to convince us that the patterns mean nothing. They're to be dismissed.

Artificial Reality

We exist largely in an artificial reality brought to you by the makers of the latest pill or injection. It's a reality where invisible forces work daily to hype fears about certain illnesses, and exaggerate the supposed benefits of treatments and cures. On the other hand, these forces have financial reasons to minimize or deny spikes in other types of illnesses that they make money treating, so that we don't seriously explore root causes. In those cases, they argue the surge is just the result of greater awareness and better diagnosis, or maybe the invention of overactive imaginations.

The hard truth is that every emerging disorder we suffer opens up new opportunities for pharmaceutical companies to earn millions or billions. Drug companies and their government partners fund the majority of scientific studies with the goal of promoting a product. Studies that could stand to truly solve our most consequential health problems aren't done if they don't ultimately advance a profitable pill or injection.

The government subsidizes food that's bad for us; allows known neurotoxins, cancer-causing chemicals, and hormone disruptors to be in the food we feed our children every day and the water we drink; and restricts practices that could make us healthier. It incentivizes the practice of medicine in a way that reduces us to a series of target numbers to be achieved through expensive pharmaceuticals and treatments, and punishes doctors who color outside the lines and practice independent patient care. Why?

Misinformation issued on a grand scale saturates the information landscape and becomes impossible to avoid. The government, federal agencies, doctors, medical associations, and cleverly disguised nonprofits backed by industry push this misinformation on the public with impunity. Medical professionals and scientists who question the dogma and offer independent research or individualized treatment to patients are held up to ridicule, punished, even yanked before medical boards to have their licenses pulled.

Even most of your doctors don't know that the medical journals they rely upon are filled with unreliable studies hopelessly tainted by drug industry interests—and that's according to some of the journal editors themselves! Studies are ghostwritten by drug companies, then held out as independent. Researchers, under pressure to publish, can buy studies with fake data on the Internet and put their names on them.

Part of the reason your doctors don't dig deeper lies in the fact that medical school curriculae and professional Continuing Medical Education classes are designed by the pharmaceutical industry too. The media make an incredible fortune selling drug ads, so their news and information arms look the other way on matters of drug safety, or even aid and abet the misinformation. Politicians gobble up contributions from the pharmaceutical industry and staff their congressional offices with drug industry personnel who make sure pesky oversight hearings on all these topics are never held. It's a scandal of mega proportions—but who is left to expose it and correct it?

In this artificial reality, success isn't measured by the good health of the population. Quite the opposite. It's measured by how many people are taking expensive drugs or getting vaccinated. It's not about preventing illnesses; it's about treating them indefinitely.

Vaccines: A Change in Definition

In the case of vaccine makers, success comes with inventing shots that can be added to the list of what's required for schoolchildren. Better yet, invent shots that the public can be convinced to get, repeatedly, for the rest of their lives. *Instant billion-dollar blockbuster!*

This has led to a questionable dynamic where the onetime standard that vaccines were required to meet—that they must be vital, safe, and effective—fell by the wayside. Instead, the government aggressively serves as promoter of dubious versions that may not be necessary, may not work very well, and come with the risk of serious side effects.

In 1975, the cost of vaccinating a child from birth to age six was $10 (in 2001 terms, adjusted for inflation). As more vaccines were added to the list, the cost ballooned to $385 in 2001. Today it's thousands of dollars. The costs are largely hidden to us since we get inoculated for free or with minimal out-of-pocket payments. But make no mistake, we're paying the bills in the form of insurance premiums, and tax dollars to state and federal programs that provide vaccines at little to no direct cost to the patient. Vaccine companies are reaping enormous profits.

Sometimes getting and keeping a vaccine on the market requires sleight of hand. The Centers for Disease Control (CDC), our premier infectious disease federal health agency, is happy to give a little help to its vaccine industry partners or, as the CDC calls them, "stakeholders." The agency's best and brightest can even adjust the veritable meaning of the word "vaccine."

The CDC used to define "vaccines" quite simply as agents that "prevent disease." But in 2021, that had to be changed. It became undeniable that Covid vaccines didn't prevent the disease (or transmission, or even illness). Logic might suggest that the Covid vaccines would have to be withdrawn from the market. After all, they didn't even meet the definition of a vaccine. Instead, the CDC quietly redefined the word "vaccine" to make the Covid shots seem successful after all.

On the CDC's vaccine web page, sometime between September 1 and 2, 2021, somebody removed a key phrase from the definition. On September 1, the CDC defined a vaccine as "a product that stimulates a person's immune system to produce immunity to a specific disease, protecting the person from that disease." But on September 2, the phrase "protecting the person from that disease" was removed, like it never even happened. Now, the CDC says, vaccines merely "stimulate the body's immune response."

Think of it. The CDC unilaterally redefined two hundred years' of the world's understanding of what constitutes a vaccine, without so much as an explanation, public discussion, hearing, or vote. Once you understand that our top, trusted medical authorities are willing to sneakily move goalposts and change meanings of words to protect a market, you're a long way to beginning to understand how deep the corruption goes.

It's one thing to be barraged by marketing to convince you to buy a shiny new car. But it's quite another to get sold a bill of goods by our trusted health experts when it comes to our most precious possession. Our increasingly elusive quest for good health has become a commodity to be bought and sold by today's snake oil salesmen and their coconspirators, but on a far grander scale.

The global demand for prescription medicine is projected to hit $1.9 trillion ($1,900,000,000,000) by 2027. These aren't necessarily drugs designed to make us well, but ones we'll "need" for life.

There's never ever been a weight loss drug that hasn't been pulled from the market for safety reasons over time. Yet each new entry to the market is heralded, applauded, and embraced by the medical establishment and a public eager for an easy answer to a complex question. Maybe the latest fad, Ozempic, will be different. It's a once-a-week injection for diabetes. But it's no accident that most of the world now knows it as a weight loss medicine. It's quickly become so popular and easy to obtain, patients without diabetes can get Ozempic with a one-time telephone consultation, or from a nurse at their favorite medspa. Never mind the side effects, including thyroid tumors, pancreatitis, vision changes, low blood sugar, gallbladder issues, kidney failure, and cancer. And if some patients report suppression of appetite so serious they become malnourished even after they stop taking Ozempic; or if others report severe paralysis of the stomach, so be it. Maybe it too will be pulled from the market someday. But in the meantime, there's money to be made.

In their defense, pharmaceutical companies are doing exactly what they were built to do: make money. The thought that they're somehow different from other multinational corporations, that they are motivated by altruism and can be trusted to be honest about the failings of their own products, is a fallacy. There's no law that requires them to put patient health ahead of profits. There's nothing that forces them to stop promoting a pill even if they secretly know it doesn't work or has dire side effects. It could be argued they have a fiduciary duty to try to downplay or even cover up negative information about their products if it could hurt their bottom line.

Our sick and broken system is the fault of politicians, federal agencies, the medical establishment, and the media. They have a far different responsibility than private drug companies. But they've allowed themselves to be so captured by commercial interests that they function as little more than an advertising arm of the pharmaceutical industry.

Looking Ahead

There may be a silver lining. The bad guys finally went too far.

With Covid: the disinformation, intolerance for dissent, shutdowns, mandates, forced or withheld medical treatment, mass firings, and attacks upon tens of thousands of scientists sparked the formation of a diverse coalition. This coalition includes a mix of liberals, conservatives, and nonpartisans. It's made up of freethinking parents, students, doctors, nurses, researchers, elected officials, and celebrities.

Many had never before questioned public health narratives or their doctors. Most had blindly supported them. But today, members of this new coalition find themselves probing widely pushed orthodoxy on Covid and beyond, rightly asking what else the media and top public health officials have misled us on.

Now, redemption from the grasp of those who seek to control our health and our lives may come through a collective awakening that's already begun.

CHAPTER 1

Down the Rabbit Hole

I blame the 9/11 terrorists for all of it. For creating a new, unexpected facet to my journalism career. For an enlightenment I wasn't seeking. For my entry into the Rabbit Hole.

On September 11, 2001, I'm driving down the George Washington Parkway in Virginia alongside the Potomac River, headed to work. Oaks and elms lining both sides of the road are still heavy with summer green, though fall is but a whisper away. The sky is a brilliant sparkling blue, the air wiped clean by a storm that rolled through the day before. But I'm running late. I'd stayed home longer than usual to watch the TV news after the first terrorist-hijacked jet crashed into the North Tower of New York's World Trade Center. It doesn't look like an accident. *The pilot wasn't trying to miss. He aimed directly into the side of the building.*

A little over fifteen minutes after the collision, I'm in my car listening to the news on the radio as plane number two flies into the World Trade Center's South Tower. My mobile phone sounds. It's my husband, Jim.

"So, definitely terrorism—not an accident," I say to him, stating the obvious.

Would the next hit be in the nation's capital? In my sixteen years of working at the Washington, DC, bureau of CBS News, we occasionally referred to the inevitability of our city being a terrorist target.

What would you do? we'd ask each other. *Stay in the city and cover the story? Or try to get out? How would CBS continue broadcasting the news out of the capital if our technical capabilities were interrupted?*

A crusty old producer at the bureau used to share his plans. When the eventual attack arrived, he said he'd get a folding chair from his office, take the elevator to the ground floor, walk out the front door of 2020 M Street NW, open the chair, sit down, bend over, and "kiss my sweet ass goodbye, because nobody's going to be able to make it out of the city."

As I round a bend on the GW Parkway, I notice a pillow of white smoke rising above the tree line ahead of me. *Holy shit! What's on fire?* It seems like seconds, not minutes, before the newsman on the radio fills in the blank. Now a plane has crashed into the Pentagon, here in the DC area.

Like millions of people hearing reports of the attacks, I have a decision to make. Either I find a spot to turn the car around, go get Sarah out of her first-grade class back in Leesburg, Virginia, and take her home, or I cross the bridge from Virginia over into DC with the understanding that I may not be able to get back out anytime soon. And the terrorists could well fly a plane into the US Capitol Building. Or the Washington Monument. Or the White House. *All I'd be able to do would be to get out that folding chair . . .*

I try calling Jim back, but mobile phones are rendered useless, apparently overloaded by frantic people all trying to make calls at once. *What a disaster our systems are when an emergency strikes!* I cross the bridge into DC.

Traffic gets so jammed in the melee from the masses trying to evacuate that people abandon their cars in the middle of the road. I do the same. We lock them up, leave our vehicles right there in the street, and hoof it the rest of the way to the office. Nobody will get out of the city on this night.

At CBS, any investigations or other stories I've been working are permanently shelved. Our news coverage will become all terrorism, all the time for the foreseeable future. It is because of this fateful turn in assignments that I'm about to stumble into something that will draw me into a shadowy underworld, one where untold dangers lurk beneath the cloak of "public health." Where misinformation reigns supreme and where government and corporations collude to corrupt science and destroy those who could expose them.

The Smallpox Vaccine Comeback

In the frenetic aftermath of 9/11, one of the most virulent diseases known to mankind becomes a key focus of concern. Smallpox hasn't been seen in the US in nearly thirty years. Now it's considered a potential bioweapon that terrorists could deploy on a massive scale. Smallpox is ultracontagious and might spread undetected for days before it's recognized.

Smallpox is transmitted through droplets coughed or sneezed into the air. People can also get infected by touching a sick person's pustules or poxes. Or even if they touch bedding that pustules touched. The virus, we're told, can kill up to one in three of those who get symptoms. Maybe with today's modern medicine it wouldn't be as fatal as in the past, but who wants to take the chance? Death comes from toxins in the blood and systemic shock. Routine vaccination against smallpox ended in 1972, when the disease was no longer a threat here. *We're sitting ducks.*

CBS assigns me to cover a big story: the planned restart of a mass smallpox vaccination program to protect against this new bioterror threat.

The topic of vaccinations is a relatively new one for me. I'd never been particularly interested in digging into medical stories, though CBS had recently assigned me to investigate dangerous prescription drugs, and I'd broken some news on that front. I quickly learned that medical scandals have similar components to other scandals I've covered. Money and greed are typical drivers. Corrupt partnerships between government and corporations are common. Well-funded propaganda efforts control the narrative. And facts are often exposed with the help of insiders blowing the whistle.

The main thing I know about vaccines is that my daughter has had all of hers. Measles, mumps, rubella, tetanus, pertussis, diphtheria, polio, the new chicken-pox shot—whatever the doctor recommended. And I've had more than my share. When you're a news correspondent in the press pool who can be called up at a moment's notice to travel with the US military, as I am at the time, you're expected to stay current on even some exotic vaccines not routinely given in the States. That's all well and good, but a few years back when my phone rang and I was dispatched to a secret foreign destination with the troops for CBS News, I was in for a surprise. I reported to Andrews Air Force Base and eagerly flashed my yellow, slightly tattered World Health Organization vaccine card at the military medical officer doing health screenings, only to have him wave it off and vaccinate me again. Turns out the military doesn't care what shots you've already had. They don't tell you what's in the syringes they're injecting you with now. And they don't keep records of it. (I would discover that fact years later after I spontaneously lost my senses of smell and taste following a battery of military shots, and my own doctor wanted to know what jabs I'd had. But when I asked, the military told me no records exist.)

Anyway, on December 10, 2002, I break the news internationally on CBS. After nearly a year of discussion and stockpiling of newly minted vaccine, the smallpox inoculation program now has a firm date to start back up in the United States! Personally, I can't wait for my daughter to be eligible to get that protection.

I quickly begin getting up to speed and reporting on the new plan, as well as the emerging safety questions. Smallpox expert Jonathan Tucker tells me the decision to give mass vaccinations is "agonizing." *Why?* "Because," he says, "it involves weighing the risk of a possible terrorist use of smallpox, which is almost impossible to quantify, against the known risks of the vaccine, which are substantial."

"The initial people are going to be test subjects or guinea pigs," adds Dr. Martin Blaser in my story for the *CBS Evening News*. He's an infectious disease specialist and chairman of the Department of Medicine at NYU Langone Medical Center.

Three days later, President George W. Bush makes it official. The first shots will go to 500,000 military troops and 450,000 frontline healthcare workers.

"As an example to military personnel who may be concerned about the vaccine's risks, Bush also . . . will receive a smallpox shot himself," reads a report from the federal Center for Infectious Disease Research and Policy. But "[g]iven the current level of threat and the inherent risks of the vaccine, we have decided not to initiate a broader vaccine program for all Americans at this time."

I also learn that the Centers for Disease Control is planning an aggressive information program to help Americans "weigh the risks" of vaccination and "let them decide for themselves." The CDC is said to be convening an expert board for real-time monitoring of adverse events reported after smallpox vaccination.

As smallpox vaccinations begin in 2003, I continue my contacts with government advisors and other medical experts. I'm developing sources whose expertise will become valuable long past the smallpox questions. I start to discover why there's so much debate about the smallpox vaccine. To summarize three key reasons:

1. The smallpox vaccine has a higher rate of serious side effects than routine vaccinations. It's estimated that out of every million people who get the shot, at least one to three of them will die. That could translate to hundreds of deaths, and thousands of life-threatening illnesses caused by the vaccine.
2. The vaccine actually makes some people contagious with smallpox for a time.
3. Smallpox vaccination can be effective *after* exposure, which reduces the need to pre-vaccinate millions.

After three months go by, only a fraction of the targeted healthcare workers have gotten the smallpox shots. They're skeptical. The government safety surveillance program has detected an unnerving trend: eleven cases of unusual heart inflammation among vaccinated military troops and three civilian deaths. The injuries get quite a bit of traction in the press.

But it's just eleven cases and three deaths among tens of thousands, I think to myself. *Why is that cause for such alarm?*

I investigate further and learn a fact that would resonate for decades to come as I cover an expanding array of medical issues and scientific scandals. The actual number of injuries from vaccines or other medicine is accepted to be manyfold higher than what gets officially reported. Each single reported adverse event, experts say, can imply anywhere from 10,000 to an astonishing 100,000 *additional* injuries that go unreported.

Back to the smallpox vaccine. With the reports of heart inflammation in early 2003, Dr. Brian Strom urges the government to slow—or stop—this restart of inoculations. At the time, Dr. Strom is head of the independent committee advising the government on the program.

"This is a toxic vaccine," Dr. Strom tells me. "We should only use it in people who need it. And we need a few weeks or months to step back and say, 'Let's replan the plans to see how many people need to get the vaccine before we continue on with it.'"

All of this sends my critical thinking skills into overdrive. *A "toxic" vaccine? Didn't the smallpox vaccine save the world, once upon a time?*

I research more about supposed side effects. I unearth patient claims about "deep vein thrombosis"—blood clots that can travel to the heart or lungs and kill. I find reports about heart attacks caused by inflammation of the heart muscle or tissue after smallpox vaccination. The "scientific community" seems to be debating whether these illnesses are truly linked to the smallpox vaccine, but some authorities believe they are. My search organically leads me to scientific literature discussing other vaccines. The combination of smallpox vaccine and the controversial anthrax vaccine, also given to military troops at the time, could magnify the risk of side effects. Until now, I'd never given much thought to how getting multiple vaccines could be riskier than getting them one at a time.

On March 25, 2003, the director of the CDC, Dr. Julie Gerberding, addresses mounting worries about the smallpox vaccine.

"We have received several reports of cardiac-related problems among the 25,000 or so people who have been vaccinated in this program," says Dr. Gerberding. "One of these patients has died, and the other is on life support . . . over the weekend we learned that an additional patient had died. . . ." But during the same teleconference, she also seems to downgrade concerns.

"I would just stress again that all of these individuals had very clear risk factors for myocardial [heart inflammation] disease. So we are not talking about people with no unknown reasons to have ischemic [blood supply] disease. And that's why we cannot necessarily ascribe any relationship of vaccination to these events. It could be entirely coincidental," says Dr. Gerberding. In fact, the starting point, Dr. Gerberding tells us, is the "coincidence" hypothesis. "I think the first hypothesis is that this is not causally related," she tells reporters of the heart deaths.

Her remarks trigger some serious cognitive dissonance on my part. When it comes to many other health hazards, a cautious approach is often taken—almost to a fault. In those cases, no risk is considered too small to require a warning. Ladders are labeled. "*Don't fall off the ladder.*" Plastic bags come with cautions. "*Suffocation risk. A plastic bag is not a toy.*" But when it comes to medicine that millions of people take, there's an opposite approach. No potential risk seems big enough to warrant concern; no proof of adverse events is convincing, even when it comes to life and death. If any alternate theory can conceivably be blamed for a person's injury after taking medicine, then that's the explanation we are to accept. Government and drugmakers work together to prevent mandatory warnings. If warnings cannot be warded off entirely, they are stalled for as long as possible. In the time span of the first danger signal and when a warning finally makes it on a medicine label or the drug gets pulled from the market, countless patients will be injured or killed.

But there's a bright side for the drug companies: they use all that extra time to sell more product to cover the cost of the inevitable court claims that will be filed against them.

Everyone is expected go along with this system, it seems.

■ ■ ■

The Curious Case of David Bloom

It's April 6, 2003, when a horrible tragedy hits close to home.

NBC journalist David Bloom, a colleague I personally know, has lost his life while deployed overseas. He'd been embedded with the military for the war in Iraq. His death isn't from a battlefield injury. It's from a blood clot that started in his leg and moved to his lungs. *Deep vein thrombosis*. He was only thirty-nine. Having recently read research about that very condition occurring after smallpox vaccine, I wonder: *Had David just gotten the smallpox shot as part of his embed with US soldiers?*

No mention of it in the press. And nobody but me seems to be making a theoretical connection. I'm starting to discover how remarkably uncurious the science press can be when it comes to puzzles unearthing medicine-induced injuries.

The wartime death of a journalist shakes the television networks. They react by calling home most of their war-"embedded" journalists, leaving only a critical few in place closest to the battle.

As for me, I've been embedded with the US Air Force at Ramstein Air Base in southwest Germany for CBS News. I'd been through abbreviated training with the troops and had gotten word that I'd be flying with a group of F-16s on combat sorties to drop bombs on Iraq on March 22, 2003. It was to be a night of "Shock and Awe" (as it was billed) to bring Iraqi leader Saddam Hussein to his knees. But when that night arrived, we waited and waited for high-up approval to actually launch our jet. We eventually got word that Turkey—our supposed ally—was refusing to give US jets permission to land for necessary refueling. Therefore we couldn't fly the mission. This was a big reason why, I was later told, Shock and Awe was far less shocking and awesome than the military originally envisioned. Turkey's role in keeping out our jets was never, to my knowledge, publicly reported. My next embed assignment, if I remember correctly, was going to be for me to parachute into northern Iraq's Kurdish territory as part of an Air Force group that would build a runway for US landings. But when David Bloom died, CBS abruptly called me back to the US.

I hadn't gotten a smallpox shot in order to embed with the Air Force. From what I understand, I would have been told if I had been given that particular jab. Also, I would know because it's a unique injection that leaves a scab on your arm. I didn't have any scabs after my most recent round of mystery military shots. David Bloom, on the other hand, had been embedded with US ground troops who typically *were* getting the smallpox shot. Hearing of David's death, I contact several sources, including another journalist who knows him. At the time, this journalist is also embedded with the US military in Iraq. He confirms that David *had* gotten the smallpox vaccine—and likely the anthrax shot too. "A one-two punch" for side effects, some would later hypothesize.

I know from my reporting so far that any death after smallpox vaccination must be reported for evaluation by the federal bodies tracking adverse events in real time. It's not a *suggestion*; it's a *mandate*. Patterns among reported injuries can reveal vaccine-induced illnesses already known or as-yet undiscovered.

But in news reports, nobody seems to mention David's vaccinations or connect them to his death—even as the government is rethinking the entire smallpox vaccine program due to mounting injuries. *Why isn't anyone picking up on it?*

I search for additional sources to confirm what I've learned and try to find out whether anyone has reported David's death to the appropriate database. The military hadn't done so—they're claiming zero deaths so far after the smallpox vaccine. *Surely, the experts wouldn't purposely omit David's death from mandatory reporting. They wouldn't be so bold as to try to cover it up—would they?*

Before I conclude that David's death hasn't been properly recorded, I need to find out if he's one of the three reported deaths of unnamed civilians. I had assumed those victims were first responders, but maybe one of them was actually David. The problem is, the government tells me it can't give out the names of any of the civilian victims, for privacy reasons. I come to learn this is a convenient catch-22 that makes investigating such medical controversies more difficult. The government is often happy to release all kinds of information if it promotes something positive about a medicine, but it guards information like a national secret when it could negatively reflect on pharmaceutical interests.

Maybe I can't get the three civilian names whose deaths are under investigation, but what if I go about it a slightly different way? The question isn't *What are the names of the civilian smallpox vaccine victims?* but rather *Is David Bloom's name among them?*

No, I'm told by a reliable source with firsthand information.

This is major news. Poor David may have died as a result of his vaccinations, but also nobody has properly reported his case for analysis.

On May 12, 2003, I break the story on the *CBS Evening News*:

> CBS News has learned of one high-profile death that hasn't yet been counted—that of NBC Correspondent David Bloom. He died of an apparent blood clot several weeks after getting both the smallpox and anthrax vaccines. Asked if an individual death that occurred within a matter of weeks of a smallpox vaccination should have been reported, [Dr. Brian Strom, head of the independent advisory committee on smallpox vaccine] said, "Yes." . . . Bloom's case would make four deaths under investigation for a possible link to the smallpox vaccine. Already considered the riskiest of its kind, the smallpox vaccine may be even more dangerous than anyone thought.

I later learn that, as a result of my reporting, someone quietly added David's death to the civilian mortality count. *How could I have been the only person in the world to make the connection? What's the true number of injuries and deaths after smallpox vaccination? What else is slipping through the cracks?*

I'm knee-deep in the Rabbit Hole.

In a normal news environment, the vaccine angle to David Bloom's story would make international news. But for the most part, the bombshell is largely ignored. Even David's own network, NBC, seems to avoid the topic. *Strange.* One could normally assume that David's family would want others to know as much as possible about what may have killed him. If NBC won't tell the cautionary tale, I figure the family will be glad to know there's a reporter who will. I reach out to them, but I am told they don't wish to talk. As far as I know to this day, David Bloom's family never publicly discussed a possible vaccine link to his blood clot. Instead, they became active in efforts to publicize the risk of deep vein thrombosis or DVT. You see, David, we were told, had a genetic predisposition for DVT. I wonder. *Who thought to examine David's DNA? And why aren't officials disclosing that a "genetic predisposition" can magnify a person's susceptibility to a particular vaccine side effect, rather than exonerate the vaccine as a cause?*

In August 2003, I reported on another sad death among the troops that happened the previous spring. Army Reservist Rachel Lacy got her military shots on March 2, a month before David Bloom. She became extremely ill and died in a matter of weeks. Only because she happened to have a civilian coroner examine the body were we able to later learn details. Lacy's father said that her body suffered a severe autoimmune reaction of some kind. "Recent smallpox and anthrax vaccination(s)" were listed as contributors to her death. It's like her internal organs were "dissolving," one official said. Yet the military failed to mention Lacy under "Noteworthy Adverse Events" in an article in the *Journal of the American Medical Association* touting its smallpox vaccine success. It claimed no deaths. "The government is covering this up and it is a doggone shame," said Rachel's father, Moses Lacy.

A later investigation by United Press International (UPI) would uncover a number of unusual deaths among troops after anthrax and smallpox shots, including cases that resembled David's. In an October 21, 2003, report entitled "Mystery Blood Clots Felling US Troops," UPI Investigations Editor Mark Benjamin tells of soldier William Jeffries, thirty-nine, who suddenly collapsed in Kuwait and died a week later. Days before Jeffries' death, his family said he was seen with "a scab on his arm from his recent smallpox vaccination." According to one assessment, Jeffries had "the largest pulmonary embolism" the doctor had ever seen.

In November of 2003, amid pressure from news reports and findings from two civilian panels, the military finally admitted vaccinations "might have led to the death" of Lacy. She'd received "smallpox, anthrax, typhoid, hepatitis B, and measles-mumps-rubella" shots all in one day and experienced "a complex set of pulmonary, neurological and other symptoms" afterward. She also had inflammation in her heart. "She died due to a severe inflammatory process affecting her lungs, findings consistent with a diagnosis of systemic lupus erythematosus (SLE) or lupus," according to the Defense Department in a statement. Military advisors concluded, "the weight of available evidence favors acceptance of a causal relationship between the immunization experience and the disease."

Coincident with my exposé on David's death, the feds were already "allowing" states to stop vaccinating civilian healthcare workers and first responders due to all the safety concerns. Only about 35,000 of the targeted one-half-million healthcare workers got the shot before the smallpox vaccine program died its unceremonious death.

In 2013, ten years after David's passing, his wife appears on NBC's *Today* show to promote National Deep Vein Thrombosis Awareness Month. "In the past ten years, we've established March as National DVT Awareness Month, and we've raised awareness by about twenty percent," she tells NBC's Matt Lauer.

Appearing alongside her on NBC is Dr. Geno Merli. Merli happens to be a paid consultant to Sanofi-Aventis, which manufactures a deep vein thrombosis therapy and a smallpox vaccine.

CHAPTER 2

Accidental Enlightenment

What I learned through reporting on the smallpox vaccine helped remove blinders I didn't realize I was wearing. With my peripheral vision fully activated, I began to identify important scientific scandals all around—stories that few journalists were covering fully or fairly. The perversion of science is so commonplace, it's fair to estimate there are thousands of real-world cases where it's costing lives.

For all of the difficult stories I've covered on scientific misconduct, there's one that occupies a special part of my brain where I cannot seem to tuck it away or forget it. I call it the Baby Oxygen study. Officially, it's titled "SUPPORT." Funded with our tax dollars, and with the participation of major establishment institutions, the story of the SUPPORT Baby Oxygen study drives home the disturbing callousness our government, public health officials, and top researchers are capable of when it comes to the most vulnerable among us. It helped me to understand, in a graphic sense, that some of the most trusted within our society are willing to commit life-threatening ethics violations. When caught, they circle the wagons and marshal forces to defend their bad acts, and take steps to cruelly silence the injured and smear those who would expose them.

It's understandable if you approach this tragic story as if it were an anomaly, because it makes so little sense on a human level. Unfortunately, it's indicative of countless examples of ethical lapses occurring on a daily basis in American medicine. In this instance, as you will see, medical authorities allegedly misled new moms into enrolling their premature newborns into a risky study. A medical watchdog group determined that some of the babies died as a result. The behavior of the medical experts involved violates the basic tenets of "informed consent," a binding principle explicit in medical ethics and law. Informed consent requires doctors and researchers to provide patients with full information about the risks of treatments or studies, so they have a full understanding before they consent or decline. But wait until you hear what happened when the Baby Oxygen study scientists got caught, and what eventually became of a top official entangled in the mess.

The Tuskegee Precedent

Modern rules for research on humans were forged after the US government's Tuskegee syphilis experiment on black men in 1932, entitled "Tuskegee Study of Untreated Syphilis in the Negro Male." For forty years, test subjects in the Tuskegee experiment weren't told they were part of a study, nor were they treated for their syphilis, even after penicillin was determined to be a cure in 1947.

An Associated Press series exposed the study horrors in 1972. An outcry led to new rules designed to prevent a repeat of the Tuskegee debacle. Now study scientists in America are required to tell participants about risks and get their written consent to take part in research. Additionally, study designs and consent forms must be approved by expert panels called Institutional Review Boards.

But decades after the Tuskegee experiment ended, the SUPPORT Baby Oxygen study would reopen painful wounds and raise serious questions about the actions at prestigious facilities such as Duke University Hospital and Yale University School of Medicine. How could they all have made, in the opinion of a government ethics watchdog, such critical errors?

■ ■ ■

The Awful Beginning

On October 11, 2006, twenty-five weeks into her pregnancy, Sharrissa Cook gave birth to a critically ill baby boy, Dreshan. Dreshan weighed in at a fragile 1 pound, 11 ounces. He lay motionless in the incubator, connected to tubes and monitors in the neonatal intensive care unit at University of Alabama at Birmingham Hospital.

"He was so tiny," Cook tells me, pointing to a photo of her infant. "I was a first-time mom. I didn't have a clue. I didn't know what to expect."

Medical personnel asked Cook, then a twenty-six-year-old single mother, to enroll little Dreshan in a study described simply as a program offering support and assistance. She readily signed on. "I remember them telling me they were a support group who would pretty much hold my hand through the developmental process," Cook tells me. She says she had no idea what they were really going to do.

In reality, the SUPPORT Baby Oxygen study was a national, government-funded experiment on 1,316 extremely premature infants whose fate may as well have rested with the flip of a coin. Other single moms who were prompted to sign up their critically ill babies at University of Alabama at Birmingham Hospital describe similar misrepresentations.

Bernita Lewis, then a twenty-two-year-old student, told me she enrolled her premature newborn, Christian, after medical personnel said it was nothing more than permission to gather statistical data such as weight and height. And Survonda Banks, then twenty-one, unemployed and on public assistance, says someone handed her a consent form on her way into an emergency C-section at twenty-eight weeks of pregnancy. Banks remembers being told only that it was a way to help her baby, Destiny.

The government-backed Baby Oxygen study was named SUPPORT, which stands for "Surfactant, Positive Airway Pressure, and Pulse Oximetry Randomized Trial." It took place at twenty-three academic institutions from 2005 through 2009 under the National Institutes of Health (NIH), part of the Department of Health and Human Services (HHS).

The three women I spoke to, all of them black, told me they never would have signed consent forms to put their fragile preemies in the SUPPORT study had they known its true nature, which they only discovered years later by happenstance. "[Dreshan] was already at a slim chance of surviving. Why would I make his chances of surviving more slim?" Cook later asks me, rhetorically. By the time I interviewed the moms, two of the children, Dreshan and Christian, were seven years old and struggling with a myriad of health problems. Little Destiny's fate was far worse. She died within three weeks of entering the study. The mothers wonder, *Did the experiment contribute to any of our children's medical problems?*

"The word 'unethical' doesn't even begin to describe the egregious and shocking deficiencies in the informed consent process for this study," says Dr. Michael Carome, an internationally recognized expert on research ethics with the Washington, DC–based consumer watchdog group Public Citizen. "Parents of the infants who were enrolled in this study were misled about its purpose."

And you paid for it. Although federal officials repeatedly claimed the government didn't know how much the SUPPORT Baby Oxygen study cost taxpayers, public documents later obtained through a Freedom of Information Act request put the figure at $20.8 million.

Oxygen Dilemma

Medical personnel routinely give supplemental oxygen to preemies with immature lungs—generally those born before twenty-five weeks. Too much oxygen can cause severe eye damage. Too little oxygen can lead to brain damage and death. The stated goal of the SUPPORT study was to precisely identify the lowest level of oxygen that would preserve vision, yet be sufficient to prevent brain damage and death. The results wouldn't help the study babies at all, but could guide treatment of future preemies.

The biggest problem, said critics, lies in how researchers sought their answers. There were key differences in how they treated newborns in the Baby Oxygen study compared with those not in the study. Oxygen levels for the study babies were maintained within a generally acceptable range. But medical personnel were prevented from the normal act of constantly adjusting oxygen levels as the preemies' conditions changed, based on their individual needs. The SUPPORT study arbitrarily assigned infants to either a high-oxygen or low-oxygen group and kept them in their randomly assigned range, despite their individual needs. And in a decision one government source says shocked even seasoned researchers when they learned of it, *the babies' oxygen monitors were intentionally altered to give false readings*. The reason? So medical staff in hospitals caring for the sick babies wouldn't be tempted to adjust the oxygen for the good of the individual baby! "Nothing in the consent form explained that the falsely reading oxygen monitors could lead to adverse decisions about care of the babies," Dr. Carome told me in an interview. And none of the consent forms mentioned the risk of death.

More of the high-oxygen babies in the study ended up with serious vision disorders. The low-oxygen preemies were more likely to die. The predictable and awful results were glossed over in the study eventually published in the *New England Journal of Medicine* in May 2010. But some observers who read through the material carefully identified obvious ethical conflicts. A complaint was filed by a father who wasn't involved in the study but read the published details and was apparently mortified.

The Office for Human Research Protections (OHRP) is an ethics body within the federal Department of Health and Human Services (HHS). It examined one of the complaints about the study and agreed that the consent process violated federal regulations. "The consent was significantly deficient," determined Dr. Jerry Menikoff, director of the OHRP ethics body. His office sent a stern letter to SUPPORT researchers on March 7, 2013, stating that their consent forms "failed to describe the reasonably foreseeable risks of blindness, neurological damage, and death." This was a bombshell. It implicated top medical institutions, such as Duke, Wake Forest, and Yale, in scientific misconduct.

What happened next is as alarming as the study itself.

The finding of an ethics breach unleashed a torrent of pushback—not against the study scientists but against the ethics watchdog. The scientists dug in hard. Dr. Waldemar Carlo, director of neonatology at the study's lead site, the University of Alabama at Birmingham Hospital, denied wrongdoing on behalf of the approximately two dozen institutions that took part. In a letter to the *New England Journal of Medicine* in May 2013, Dr. Carlo said they did nothing wrong because the study babies' oxygen levels were kept in acceptable ranges. He wrote: "Our consent forms were conscientiously drafted according to the Code of Federal Regulations and were based on the best available evidence."

Inside the government, officials began heated efforts to attack and marginalize their own ethics watchdog. And forty-five research scholars and ethicists got together to sign a letter urging Dr. Menikoff to withdraw his determination of an ethics violation. "We believe that this conclusion was a substantive error and will have adverse implications for future research," they wrote.

The pressure tactics worked. Less than six months after Dr. Menikoff's original determination of ethics violations in the SUPPORT Baby Oxygen study, he backed down. His office confirmed the consent forms were deficient but formally suspended all corrective action or punishment saying the researchers meant no harm and may have been confused.

In response, Dr. Carome told me, "There's no doubt in my mind that intense political pressure was brought to bear" from an academic research establishment that is dependent on the government, and that senior leadership at HHS "bowed" to it. Dr. Carome is familiar with agency politics. He used to work in the same government ethics office involved, the OHRP. A spokesman at HHS, which oversees OHRP, would not directly answer when asked whether senior government officials improperly intervened in the case to suspend corrective action.

Internal government emails later obtained through the Freedom of Information Act fueled suspicions of political interference. They show senior National Institutes of Health officials trying to tamp down the ethics probe while orchestrating a defensive commentary published in the *New England Journal of Medicine*. NIH officials "launched an aggressive campaign to undermine" the investigation into the SUPPORT Baby Oxygen study—"And regrettably found several willing partners for this campaign at the highest levels" of HHS, according to one watchdog.

Nonetheless, the federal government did come under enough pressure that it agreed to convene a big public meeting to examine the whole controversy. Hundreds of researchers and academics from around the globe gathered in person or via teleconference on August 28, 2013, to address the supposed confusion surrounding what constitutes "informed consent" in a study like SUPPORT. Even back then, I realized it was rare for esteemed researchers to be called on the carpet for unethical behavior. So I was anxious to hear how accountability would come.

My cameraman and I showed up the day of the conference at the Great Hall of the Hubert H. Humphrey Building near the US Capitol, where Health and Human Services (HHS) is headquartered. My plan was to sit in the audience, listen, and report on the event. I didn't see any other reporters there. HHS officials controlled access to the meeting like Fort Knox. I had to register, show ID, sign in, and was required to have a government escort even to visit the bathroom. This meeting was to be the first step in the government's effort to clarify and draft new guidance on the consent process for human research. But what I heard was startling.

Instead of officials expressing regret, instead of holding discussions about accountability and how to avoid a repeat, about half of the academics who spoke defended the SUPPORT study's deficient consent process. Not only that, some actually used the forum to make the case to loosen up on "informed consent." After all, they argued, when people are told all of a study's true risks, they're unlikely to take part, and the greater good won't be served. I had to do a mental reset to see if I was hearing correctly. They seemed to unapologetically argue for the right to withhold safety information from study participants—to lie and trick them—because if the truth were told, nobody in their right mind would agree to the risks.

Dr. John Lantos, director of pediatric bioethics at one of the study sites, Children's Mercy Hospitals and Clinics of Kansas City, Missouri, appeared to be on the side of telling patients less, not more. He said consent forms that make it sound like "death lurked at every corner" are counterproductive. "[T]hey are not empowering people to make informed choices, they are scaring them into making uninformed ones," he insisted. In other words, he seemed to be arguing that the more truthful information you give people, the more uninformed they supposedly are. The less you tell them, the more informed they are.

Dr. Alan Guttmacher also defended the study. He led the branch of NIH that approved and funded SUPPORT, the Eunice Kennedy Shriver National Institute of Child Health and Human Development's Neonatal Research Network. Dr. Guttmacher argued that the study added to the body of knowledge and had already helped other preemies. "We stand by this study as it was conducted and look for ways to do research even better, if there is a better way to do it, in the future," he said, apparently refusing to admit the possibility that something had gone terribly wrong. Dr. Robert Califf, vice chancellor for clinical research at Duke University School of Medicine, also part of the experiment, urged observers to dial back the rhetoric. "Sensational claims of calling people unethical, I think, really detract from the serious discussion that needs to occur," he said.

The discussion was academic until Carrie and Shawn Pratt of West Virginia took to the podium at the meeting. They were carrying their pretty, six-year-old daughter, Dagen, who was wearing a sundress and ponytails, but looked fragile and thin in leg braces. In May 2007, the Pratts agreed to sign up Dagen, a severely premature newborn, for the SUPPORT Baby Oxygen study at Duke. Like other parents, the Pratts say they were told that researchers were simply gathering information to help other children. "We never understood the study to be based on manipulating her oxygen level to meet [researchers'] needs," says Carrie Pratt. Later confronted with the reality, the Pratts were shell-shocked. They had already lost a preemie son four years before Dagen was born. They say they can't understand why medical professionals would have suggested enrolling their frail, newborn daughter in an experiment that could put her at further risk.

"When you have a small child, a micro-preemie, on a ventilator with see-through skin and fighting for her life . . . it is the most humbling, sad experience of your life," Carrie Pratt said. "So, of course we would agree to participate in a study if it meant collecting info or data to help someone else. But certainly not at the expense of our daughter."

The Pratts say that today, they wonder whether the study ultimately contributed to Dagen's health issues. She suffered multiple incidences of collapsed lungs, breathing problems, and other life-threatening conditions. Diagnosed with retinopathy, Dagen had to have laser eye surgery when she was two months old. She has cerebral palsy and often wears orthotics on both legs. "Do we blame SUPPORT?" Dagen's mother asks without expecting an answer.

Of learning that her baby had actually been part of the Baby Oxygen experiment, Sharrissa Cook told me, "That's more like playing Russian roulette to me. There's no way I would say, 'You could give my baby whatever [oxygen] you want him to have,' as opposed to what he needs." Dr. Carome agrees that it's "highly likely" that many, if not most, parents would have refused to enroll their babies in the SUPPORT study had they been "appropriately informed about the nature of the research and its risks." He told me, "Some experiments maybe just can't be done."

An interesting footnote: Dr. Guttmacher, who led the branch of NIH that approved and funded the SUPPORT Baby Oxygen study, retired not long after defending it. At that time, he announced plans to remain active in "issues of children's well-being." And the whole scandal didn't seem to sidetrack the career of Dr. Califf, the Duke vice chancellor whose institution allegedly misled the Pratt family. As of this writing, he's now the head of the Food and Drug Administration (FDA).

In early 2024 came a stunning subversion of long-standing tenets of ethical human research. The FDA issued rules that clinical researchers no longer have to get informed consent from people they experiment on when they decide their research poses "no more than minimal risk." The FDA says the new rules make it possible for scientists to conduct research that might not otherwise have been able to recruit enough consenting test subjects. It looks like the dreams of those researchers at the NIH conference a decade earlier have finally been realized. Instead of the medical establishment collectively agreeing that informed consent is a sacred and irrevocable necessity of human research, it has decided to back off the protections. The obvious problem, as we've seen, is that the new rules rely on the honesty and judgment of the very people who have proven arguably dishonest and unethical in the past. And if study risks are truly minimal, why would telling test subjects the truth about them scare them away? Don't patients have the right to know what could happen to them if they agree to be in a study? The government now says no. The lessons of the Tuskegee experiments and Baby Oxygen study are relegated to the memory hole.

But there has been a steady stream of new ones.

CHAPTER 3

Mind Games

In 2021, new hope for six million Americans suffering from the dreaded form of dementia known as Alzheimer's disease was offered in the form of a shot every four weeks. It was a drug called aducanumab, brand name Aduhelm, billed as the first treatment to attack the root cause of Alzheimer's. Never has there been a more controversial entry to market for a medicine.

In March 2019, two large clinical studies on thousands of people were underway when the maker of the drug, Biogen, announced it was halting both of them. That's because an interim analysis showed it would be "futile" to continue. No drug company wants to walk away from the chance to make billions, but the futility finding was an admission that the early study results were already so disappointing, the end result would be that the drugs simply don't work. In other words, it's futile to continue. But that wasn't the end of the road for aducanumab. And that's what makes this story so bizarre.

Nine months after Biogen stopped the studies, it made a stark reversal. Company officials now claimed they reexamined the study data in close collaboration with the FDA and found some evidence that aducanumab works, after all. *Odd that they said the opposite months before.* So now they intended to proceed to apply for FDA approval. But all they had were the two studies never completed.

I spoke with Dr. Michael Carome, who leads the Health Research

Group at the watchdog Public Citizen and closely followed the controversy. I asked, "How common is it that a drug company would present findings from studies that were never finished and try to use that to get approval of a drug?"

"It's uncommon," Dr. Carome replies. "And it's certainly uncommon for that type of evidence to be the basis for an eventual FDA approval."

Nonetheless, on November 6, 2020, the FDA convenes a meeting of its outside advisors to consider whether aducanumab should be approved for Alzheimer's. The FDA isn't required to take the advice of the medical experts serving on its advisory committees, but usually does. At this particular meeting, the best selling points that Biogen's Samantha Budd Haeberlein can come up with to present are modest. She admits the drug won't cure patients or even stop them from declining. But she says it might, possibly, slow disease progress. She tells the FDA advisors: "[W]e can conclude that the benefit-risk profile for aducanumab is favorable and potentially prolongs patients' independence by several months, even a few years, as demonstrated in our long-term study. This matters for the patients, their loved ones, and society."

Then something very unusual happens. Instead of the FDA presenting a separate, independent analysis like it usually does, it offers a single joint briefing document put together in partnership with Biogen. Speaking for the FDA is the head of the Neuroscience Office, Dr. Billy Dunn. He gives what observers would later call an inappropriately one-sided, positive review of aducanumab. He refers to one of the studies that ended in futility as a "home run." *Now it's a "home run"?*

The FDA's enthusiastic endorsement of a medicine that had failed miserably in the test phase raises a lot of eyebrows, especially because there isn't even one completed, large study normally required for a drug to get approved.

"That biased review really was so striking that it led to very harsh criticism by the [FDA] committee members," Dr. Carome later tells me. "They were especially harsh towards the FDA. They criticized the FDA for its review, for its working collaboratively with Biogen to write these analyses. And the outcome of the analyses, [the committee members] said, was one-sided—that '[the FDA was] cherry picking data [and] used an approach that was statistically and scientifically inappropriate.' It was among the harshest criticisms I've ever seen of the agency at any prior advisory committee meeting."

One of the FDA advisors who speaks out at the meeting is Dr. Scott Emerson. He's apparently as stunned as Dr. Carome. "I was very, very, very disturbed by some of the analyses," Dr. Emerson says, condemning the FDA's close collaboration with Biogen.

Adding to the controversy is aducanumab's enormous price tag. Biogen anticipates charging about $56,000 per year. Calculating ten million hypothetical Alzheimer's patients, it adds up to a mind-boggling half-trillion dollars every year, with much of the cost presumably borne by taxpayers through Medicare.

"That could lead to extraordinary economic impacts on our healthcare system," notes Dr. Carome. "It could bankrupt the Medicare program, bankrupt patients in terms of the copays that they would still need to pay. And if the drug doesn't work, at a great cost, it leads to a lot of false hope for millions of patients and their families."

At the end of the contentious meeting in November 2020, the FDA advisors vote. None of them says aducanumab should be approved. Ten of the eleven members determine there's not sufficient evidence. The eleventh member votes "uncertain." They say the FDA must require a large, randomized placebo-controlled trial—one that's actually completed—before considering approval.

What does the FDA do? It ignores its advisors and approves aducanumab.

In reporting on the controversy for my television news program, I contact Dr. Caleb Alexander. He's one of the FDA advisors who voted against aducanumab. He's an epidemiologist, internist, and professor at the Johns Hopkins Bloomberg School of Public Health.

"There was near consensus among the advisory panel that there just wasn't the evidence to warrant approval," Dr. Alexander confirms.

"Did [the FDA] really want advice from the advisors?" I ask.

"Well, it's a fair question," he replies. "And in fact, three advisory committee members . . . resigned in protest of the FDA's decision."

One of the advisors who resigned was Harvard professor of medicine Dr. Aaron Kesselheim. He wrote FDA commissioner Janet Woodcock that aducanumab was "probably the worst drug approval decision in recent US history."

Dr. Carome's watchdog group sent a series of letters to the Inspector General asking for investigations into what it called the "unprecedented close collaboration" between the FDA and Biogen that "dangerously compromised the independence and objectivity of senior staff and clinical reviewers." The FDA defended itself by insisting its interactions with drug companies are "critically important to drug development [and] essential to set clear goals and expectations . . . the absence of these interactions would dramatically delay the availability of effective drugs for patients who need them."

One of the most disputed approvals of any drug was followed by a tumultuous period once it hit the market. Investigations were launched in the House of Representatives, the Office of the Inspector General, and inside the FDA. Of course, like most government probes into the pharmaceutical industry, nothing much has come of any of it as of this writing. The FDA determined its interactions with Biogen were perfectly appropriate. A scathing report from two committees in the House of Representatives in December 2022 concluded the FDA failed to follow its own guidance and internal practices with aducanumab, and that the approval process was "rife with irregularities," but nobody was held accountable.

Meantime, questions began percolating about advertising for aducanumab. The Alzheimer's Association, consulted for advice by millions of people, plugged the brand name, Aduhelm, on its website after receiving about $2.8 million from Biogen and its partner in recent years. A promotional video on the site featured a doctor who's a paid consultant and speaker for Biogen and worked on one of the clinical studies.

I asked the Alzheimer's Association whether its advice could be biased by payments from the company that makes Aduhelm. A spokesman answered by saying that the money it got from Biogen is only a fractional percentage of the group's total revenue, and that contributions from companies have no impact on decision-making, which, they say, is based on science.

Further stoking the controversy is whether Aduhelm could be dangerous.

"There's an anecdotal report of someone who died after experiencing brain edema, which is a not uncommon side effect of [Aduhelm]," says Dr. Alexander, the FDA advisor who voted against approval. In the early months of the drug being marketed, at least four deaths were reported among treated patients. Biogen responds to questions about that by telling me it hadn't been able to establish the cause of death for one of the victims, a seventy-five-year-old Canadian woman, and that the three other fatalities "have not been attributed to treatment with Aduhelm." Yes, you're reading correctly—it's often left to the drug companies to determine whether their own drugs are responsible for deaths.

The whole example of Aduhelm is so outrageous, even the pharma-friendly medical establishment was moved to speak out. The American Academy of Neurology issued a statement saying, "[Aduhelm] is a high-cost drug that was approved by the FDA without convincing evidence of benefits and with known harms," including "[r]isks of cerebral edema [swelling], hemorrhage [bleeding], and hospitalization." The group goes on to criticize the FDA, stating that its "recent decisions indicate a lowering of the standards of scientific evidence used for drug approvals, which will require clinicians to scrutinize approved medications more carefully." The following month, the *British Medical Journal* reported that more than 40 percent of patients in a high-dose study of aducanumab suffered brain swelling or bleeding. Within weeks, the European Medicines Agency (EMA) voted not to approve it.

In the US, critics are heartened by one thing: the exorbitant price tag kept Aduhelm from being widely prescribed. It had entered the market as one of the most expensive drugs in existence. When it didn't sell, Biogen quickly cut the price in half to $28,200. But that still proved to be too high for most. The drug's only chance to survive would be if Medicare agreed to cover the cost for elderly patients with taxpayer money. But Medicare issued a death knell, saying it would only pay for Aduhelm under strictly limited circumstances.

Biogen declines my interview requests but says it "believes that patients, in consultation with their physicians, should be free to make informed decisions" about taking Aduhelm, "[p]atient safety is Biogen's highest priority," and "the potential benefits continue to outweigh potential risks."

On January 31, 2024, Biogen announces it's discontinuing Aduhelm. The loss doesn't hurt the company as much as it might have—because it had already managed to get the FDA to approve a similar, expensive drug, under equally criticized circumstances.

Aduhelm 2.0

In January 2023, under the category of "Lessons Not Learned," the FDA again grants accelerated approval to a second Alzheimer's drug, Leqembi, whose generic name is lecanemab. It's also made by Biogen, was approved amid a similar controversy, and is priced about the same: $26,500. This example shows how undaunted FDA officials are by the threat of oversight. The Aduhelm controversy is still very much alive at the time. But they know nothing serious will come of it.

Like Aduhelm, Leqembi doesn't claim to cure, dial back, or stop the progress of Alzheimer's. It may, possibly, slightly slow down decline. But how would you ever really know for sure? The anecdotal evidence isn't exactly awe-inspiring. Seventy-year-old Susan Bell was enrolled in one study of Leqembi. After four years, it didn't manage to stop the progress of the disease. "There has been, certainly, some [continuing] degradation in her cognitive powers and so forth," says her husband, Ken, in a story for NPR. But they still manage to find a theoretical silver lining. "Susan's decline has been relatively slow," NPR continues. "The couple are still able to travel and play golf, which could signal that the drug is working. 'We don't have enough experience, like the medical folks do, to know what would have happened' without the drug, Ken says."

Still, NPR reports optimistically that the Alzheimer's patient, Susan, "thinks other people in the early stages of Alzheimer's should try Leqembi. 'I would tell them, Go for it, she says. Because you really don't have anything to lose.'" *Never mind that nagging risk of serious side effects and the $26,500 yearly cost.*

Susan's uninspiring results should be of no surprise. As with Aduhelm, the potential benefit of Leqembi, if it works, is so slight, the difference would not be detectable by the patient, his family, or caregivers, says the watchdog group Public Citizen. "People are unlikely to perceive any real alteration in cognitive functioning," agrees Dr. Alberto Espay, a professor of neurology at the University of Cincinnati College of Medicine.

Regardless, Medicare agrees to cover the cost of Leqembi. In other words, powerful people in our government have been convinced to give the green light to using taxpayer money to pay the exorbitant price for a hastily approved drug that boasts only marginal potential benefits and comes with serious possible risks.

It's not long before patients are being sold on the most optimistic view of the possibilities. *Did you hear, there's a new drug that cures Alzheimer's?* a friend asks me. She proceeds to tell me how lucky she feels that one of her parents, as well as an aunt, are getting the treatment. I don't have the heart to go into the background on what I've learned.

One postscript. Dr. Billy Dunn left the FDA after his involvement in the questioned Alzheimer's drug approvals and amid outside calls for his resignation. According to a website, he quickly found a soft landing within the industry he'd regulated. He became director of a company called Prothena, which is developing the same types of brain drugs he'd helped approve at the FDA.

Market Making

America's expensive drugs wouldn't be as easy to sell to patients if pharmaceutical ads were banned from TV like they are in most every other country. But twenty-five years ago, the pharmaceutical industry and national media companies joined forces to lobby for the US to become one of the few places in the world that allow prescription drug ads on television. It was as if they'd made a deal with the devil.

The deluge of TV drug ads began around 1997. By 2020, the pharmaceutical industry accounted for 75 percent of all televised ad spending in the US, adding up to about $4.58 billion. Written with all of the zeroes, that's $4,580,000,000.00.

This financial relationship between medicine and media creates an inevitable conflict. Media companies are beholden to the drug industry that provides so much of their wealth. As a result, media executives are all too willing to censor programs, news reporting, and information that could threaten their cash cows. We're no longer viewers, readers, and users for the media companies to inform and entertain. To them we are reduced to a commodity to be leveraged into profits. The more of us who become attracted to a media source, the more that source can charge for drug companies to advertise to us. The more money drug companies pay for advertising, the more control they get over what media sources tell us about medicine.

This dynamic very much aligns the media with American political figures who also rely on and are influenced by Big Pharma money. They too serve the industry's interests rather than those of their constituents. In 2023, not counting direct campaign contributions, the pharmaceutical and healthcare industry reportedly spent more than $497 million lobbying to influence politicians and federal agencies.

Looking at a few of the top players, the Pharmaceutical Research and Manufacturers of America, PhRMA, spent $27.6 million on lobbying in 2023 alone. Pharmaceutical Care Management Association, the big industry group for pharmaceutical and insurance middlemen called PBAs, laid out $15.4 million for lobbying during the same year. Pfizer, Amgen, Merck, Roche, Eli Lily, Bristol-Myers Squibb, and Gilead Sciences spent between $8.3 million and $14.3 million each lobbying to get what they want. Those are just a few. What is it these drug industry interests want? Many desires are on the table. It could be they want to increase the chances that the government will agree to pay lucrative prices for their star products through Medicare and Medicaid, insurance for the elderly and poor. It could be that the pharmaceutical companies are paying for access so that their lobbyists are invited to help write congressional bills to their benefit—and deep-six provisions that could hurt their bottom line.

Dr. Carome of Public Citizen has studied the many conflicts generated by all the TV drug ads. "[They advertise] the newest drugs, so we often know the least about their safety," Dr. Carome tells me. "And often there are older alternatives that may be equally effective and safer. But because those drugs aren't advertised, because the generic drug industry doesn't do this type of advertising, it can [hurt] the public health overall."

Taking drugs as prescribed is a major cause of death. It's said to be a factor in more than 300,000 fatalities a year in the US and Europe. The FDA tried to mitigate criticism over allowing drug ads by requiring the commercials to clearly disclose some risks. But after watching a typical commercial, viewers are likely to come away understanding a drug's benefits, and not much about its harms.

One infamous case involves Bayer's birth control pill Yaz. A catchy Yaz commercial featured women singing "We're not gonna take it" while hitting balloons labeled with words associated with premenstrual syndrome (PMS) such as "moodiness" and "irritability." They were figuratively batting down PMS with Yaz. The problem is, Yaz wasn't approved to treat PMS and shouldn't have been promoted for that. The FDA ultimately slapped Yaz ads with multiple violations for minimizing risks and overstating benefits. Bayer ran a $20 million corrective campaign but had already made untold profits. And even after getting caught once, Bayer later produced more Yaz commercials that failed to disclose *any* risks—including heart attack, stroke, gallbladder disease, blood clots, and death, according to the FDA.

In Canada, in a five-year period (between late 2007 and February 2013) there were reports of twenty-three girls or women dying while taking Yaz or a related drug, Yasmin. Fatalities included a fourteen-year-old girl and at least eight other teens. They primarily died from blood clots in the lungs or brain. In 2011 the FDA forced Bayer to add a warning about blood clots to Yaz, finding that Yaz was more likely than other oral pills to cause clots.

Dr. Carome says when ads do include risks, they're often presented in a way people aren't likely to remember. "There are certainly advertisements where they're reading the risk information, but what's being portrayed is smiling people [or] interesting cartoon characters. And those are classic examples of distracting information." Some commercials for the diabetes medicine Toujeo fit that description. The FDA warned Toujeo's maker, Sanofi, about showing distracting, upbeat images while Toujeo's risks were being presented. The FDA also dinged ads for Pfizer's menopause drug Estring for failing to disclose "any" risks—including uterine cancer.

Another case of misleading drug commercials involves the cholesterol-lowering drug Lipitor. At one time, Lipitor was the best-selling drug in the world, generating $12 billion a year for Pfizer. The famed Dr. Robert Jarvik, inventor of an artificial heart, got a $1.3 million contract to be pitchman for Lipitor. The FDA called the commercials misleading because Dr. Jarvik isn't a cardiologist, and never treated patients or prescribed drugs. Pfizer eventually pulled the ads. But a lot of Lipitor had been sold.

The FDA isn't really serious about stopping abuses. If it were, it would impose penalties that have a true deterrent effect. As it stands, companies only have to use a bit of their profits to pay any fines. So it's worthwhile for them to skirt rules in order to pump up sales.

"Sometimes [the FDA] may find an ad that violates the regulations, and they'll write a warning letter to the company, and the company will pull the ad," says Dr. Carome. "But at that point, the ad may have been out for many weeks or months, and patients may have already been misled by it." Surprisingly, considering all of this, the FDA is said to be looking at allowing drug companies to reduce the amount of risk information they're required to present in ads.

In speaking about TV drug commercials in general, I ask Dr. Carome, "If you could wave a magic wand, what would you have the FDA do?"

"If I could wave a magic wand," he tells me, "we would no longer allow this type of direct consumer advertising at all."

"Are TV drug ads in the US ever going to go away, do you think?" I ask.

"Unfortunately, probably not in our lifetime," he says.

Study Secrets

Of course the pharmaceutical companies must have their drugs approved by the FDA before they can advertise them and get them into our medicine cabinets. This usually requires convincing the FDA that the medicine has passed a rigorous set of safety and effectiveness studies. But when companies or their government partners try to keep secrets about the studies, you can bet something's amiss.

The FDA worked to block the timely release of documents from Pfizer's Covid-19 vaccine studies even though we own the information. Why do I say it belongs to us? We helped pay for it. Our taxpayer-funded federal agencies gave taxpayer money to the vaccine makers, gave the vaccines the green light, and pressured millions of people to get injected. Information about the studies is critical to us being able to evaluate the safety, necessity, and effectiveness of an experimental medical treatment that was ultimately given to more than 5.5 billion people. We own the information and we have the right to see it.

Patient advocates sued the FDA to pry some of the Pfizer documents loose from the tight grip of the minders. The FDA tried to obstruct by asking the court to drag out release of material slowly over the next seventy-five years. But the court ordered some of the documents to be released. They revealed that Covid vaccine test subjects came down with multiple illnesses during the studies, foreshadowing some of what might happen in the general population. They include:

- A woman who suffered worsening of acute asthma within two months of the second shot
- A girl who developed deep vein thrombosis, or blood clotting, also within two months of her second shot
- A young patient with no past medical problems who suffered life-threatening myocardial infarction, or heart attack, after a second shot
- A woman who suffered a heart attack and died on November 4, 2020, after her second shot

- Two other heart attack deaths within two to three months of the shots

Pfizer wrote off the illnesses in the studies as unrelated to the Covid vaccine, and the public was none the wiser, according to the law firm Siri & Glimstad, which filed the Freedom of Information Act lawsuit.

It's grown exceedingly common that when patients get sick during a study, instead of the drug company considering the illness to be a possible side effect—which is what should be the response—they seek to explain it away. They blame anything other than the experimental medicine.

Another blatant example of this twisting of science can be found in a May 2023 study to look at whether serious neurological, or brain and nerve, disorders were connected to Covid-19 vaccines. The study was entitled "Observational Study of Patients Hospitalized with Neurologic Events After SARS-CoV-2 Vaccination." It was published in *Neurology Clinical Practice.*

The first problem I see when reviewing the study is that, although some side effects don't surface until months or years after a medicine is taken, the study scientists drew their conclusions based on a mere six-week period. They looked at only 138 people hospitalized after a Covid vaccination, and a limited number of neurological conditions, including stroke or blood clots, encephalopathy or brain damage, seizure, and intracranial bleeding.

But what really captures my attention is the study's nonsensical conclusion. It states that since all 138 vaccinated, hospitalized patients had "risk factors" or "established causes" for their neurologic illnesses, such as high blood pressure for stroke victims, *this proves the Covid vaccines are safe.* "All cases in this study were determined to have at least 1 risk factor and/or known etiology accounting for their neurologic syndromes. Our comprehensive clinical review of these cases supports the safety of mRNA COVID-19 vaccines," reads the study discussion.

You don't have to be a scientist to detect a serious flaw in their reasoning. It's like claiming that an old person who falls down the stairs and breaks a hip—was injured by being old, and it had nothing to do with the fall down the stairs. Having high blood pressure to begin with doesn't mean if you have a stroke after Covid vaccine, you can automatically rule out the vaccine as having an impact. In fact, you should immediately ask whether the vaccine might prove riskier to people with preexisting vulnerabilities.

Surely even a novice scientist should know this. So why did this ridiculous study get published? It looks suspiciously as if someone is trying to dispel growing safety concerns about the vaccines. I decide to find out who.

I learn that the study was conducted at Columbia University Irving Medical Center and New York–Presbyterian Hospital in New York City. It was funded by taxpayer money through the CDC. I email the primary study author, Dr. Kiran Thakur: "The study seems to imply that because people who suffered certain neurological events shortly after Covid vaccination had risk factors, it exonerates the vaccines from blame. But did the authors consider that people with existing risk factors could be at greater risk for vaccine adverse events?"

Instead of answering the question, Dr. Thakur replies, "Can you clarify the purpose of your questions (to be published, personal inquiry or otherwise)." When I reply that her responses might be published, she goes dark on me. When I persist in asking her to respond, she finally answers: "Declining, thank you." Why isn't a legitimate scientist happy to answer a simple question about her work? *What's the big secret?*

Reaching a dead end with Dr. Thakur, I query the medical journal's editorial staff. They loop me back to Dr. Thakur, saying only she can answer my questions. *Shouldn't the journal be asking the same questions?*

Next I turn to Columbia University. I ask to see the study materials and related communications. I want to learn *Who was behind this study, and did the peer reviewers or anybody else flag the obvious flaws?* It's a reasonable request because we, the public, funded the research and own the information. Besides, a basic tenet of scientific research dictates that there should be transparency in data and all aspects of studies. In fact, a study isn't considered legitimate unless the data is available so that it can be verified and replicated by others with the same results.

But Columbia University stalls in responding to my emails. So I file a formal Freedom of Information Act (FOIA) request for the material. More time passes, and Columbia informs me that it's a private institution and it doesn't have to follow Freedom of Information Act law. I appeal on the basis of scientific transparency. *Why does Columbia want to keep details of an important, publicly funded study secret? Isn't that contrary to tenets of sound science?* My appeal falls on deaf ears. University officials tell me they'll only respond to validly issued and served subpoenas or court orders, and that "[s]ubpoenas to the University must be served on the Office of the General Counsel."

Think of the audacity. A private university can take our tax money for a study, then refuse to answer questions about it because they're a private university. To me it looks like the CDC can legally launder taxpayer dollars to third parties to produce what amounts to propaganda, then cover their tracks under a shroud of secrecy.

Next, I decide to file a FOIA request directly with the CDC, which is undeniably subject to the Freedom of Information Act. However, I know from experience that federal agencies spin the FOIA process into a tool to obfuscate. They rarely follow the provisions requiring them to turn over materials within twenty working days. And punishment for their violations is virtually nonexistent.

Sure enough, the CDC sits on my FOIA request for forty-two days before emailing to let me know they haven't yet begun processing my request. They say I need to be much more specific, or they won't consider responding. This raises one of the newer tricks federal agencies use to make it tougher for us to access information we own. They require FOIA requests to be impossibly precise. In the past, it was enough for a requester to provide a topic and date range. Agencies would search computer records using keywords. But now they claim they can't do that.

The CDC FOIA officers now demand that I somehow discover and present them with names of each specific, archaic department and subdepartment that should be searched and the title of any documents I'm looking for. They further insist I provide names and titles of each person within those departments whose email accounts should be searched. And I must give them the number of the grant that awarded the taxpayer funds for the study. Problem is, I have no way to know any of that. The grant number was strangely omitted from the published study, and I have no clue how I would find names of the people who might have records, or what departments they work in. That's a key part of what the FOIA response would reveal. Using these avoidance tactics, a federal agency can heighten their odds of keeping public documents secret.

Havana Syndrome

Another instance of questionable application of science can be found in the case of "Havana Syndrome." Havana Syndrome is the name given to a mysterious and alarming illness said to have impacted more than 1,500 US government employees in Cuba and around the world starting in 2016. As you'll see, when there's a science-related story that certain influential people decide to back, no matter how unfounded or unproven it is, it can dominate the media and political landscape in an earth-shattering way. The entire Havana Syndrome scandal became enveloped in politics, bad science, and propaganda.

Supposed victims of Havana Syndrome reported headaches, dizziness, and a loud sound in their ears. Rumors of cases spread, with help from some sources inside the US government. Media reports and experts fueled speculation, though without evidence, that Russia or China had deployed some sort of super-duper-secret, previously unknown, new weaponry against American diplomats and intelligence officers.

As the news continued to circulate, Congress passed the "Helping American Victims Afflicted by Neurological Attacks Act" in 2021: the HAVANA Act, for short. It aimed to compensate State Department staffers with "confirmed brain injury from the disease"—a disease never scientifically proven to exist in the first place. "Victims" became eligible for more than $180,000 in tax dollars each.

In 2022, following six years of hype and speculation, the US intelligence community issued a report that appeared to poke fatal holes in the far-fetched theory behind Havana Syndrome. It concluded that most of the cases could be explained by natural causes or ordinary stress. That was followed, in 2023, by a US intelligence community assessment of reports in over ninety countries. It finally acknowledged there's no evidence to support the theory that Havana Syndrome was caused by hostile foreign powers deploying some type of secret weapon.

In July 2023, I manage to get a press visa for a visit to Cuba for my television news program *Full Measure*. After my arrival in Havana, I head to the offices of Dr. Mitchell Joseph Valdés-Sosa, director of the Cuban Center for Neuroscience. Valdés-Sosa led Cuba's investigation into Havana Syndrome. His findings were released years before the US assessment but fell on deaf ears at the time, he says, despite the fact that every bit of rational science undercut the Havana Syndrome theories from the start.

Valdés-Sosa greets me in his personal office on the second floor of the Neuroscience Center in Playa. It's north of 100 degrees outside and feels about that hot in his office. A tired room air conditioner tries its best, but manages only to wheeze out a lukewarm breeze. Valdés-Sosa sits behind his wooden desk and gestures for me to sit across from him. We chat while my cameraman sets up for an on-camera interview that we will conduct in an adjacent building that houses the center's MRI machine. Valdés-Sosa seems eager to provide his view of the scandal to an American journalist sincerely interested in hearing it. We make our way to the MRI building and start the interview.

"I think this is a tremendous story," he tells me. "Not the story that originally many people saw, you know, Flash-Gordon-mysterious-kind-of-weapon-hunting-down-US-citizens-around-the-world. I mean, that's immediately a headline story. Right? But that's not true."

His voice becomes softer, and he leans toward me a little for emphasis. "But the other story is just as interesting. It is, *How could science be sidelined for seven years?*" He sits back again and mops sweat off his face with a white handkerchief, only to have it glisten back to the surface within seconds. A floor fan is trying to pump in some slightly less hot air from another room but it isn't helping.

"The narrative that was accepted very widely," he continues, "is that a group of diplomats or spies that were in the American embassy had been attacked by some mysterious energy weapon that produced a new syndrome, a new disorder, with brain damage, with damage to the hearing. And that this same kind of attack later generalized to other countries . . . including the US."

I ask what he thought when he first got word of the supposed "new disorder." He says he was on a scientific trip to Mexico at the time.

"Some colleagues showed me a newspaper saying that people had suffered sonic attacks," he says. "And immediately I thought, 'This is impossible,' because you can't harm people with sound weapons without many people learning about this. There are some weapons for crowd dispersal, but they're so loud and, let's say, impossible to ignore, that everybody around hears about it. This mysterious idea of mysterious attacks produced by sound weapons was, in my opinion, absurd."

"At first did Cuban authorities treat this as though there might be something to it?" I ask.

"Yes. Yes," says Valdés-Sosa. So when he returned from Mexico to Havana, he says he immediately contacted Cuba's Academy of Sciences. A team of national experts was created to examine the phenomenon. It included doctors from different fields—ear specialists, epidemiologists, sound engineers, physicists, and neuroscientists.

"The first reaction . . . is that this had to be looked into very seriously. That it could happen, that somebody attacked. And they were going to look into it," he tells me. "The problem is, as more and more research was carried out from the very first moment, things didn't make sense. And the Cuban police reached the conclusion that there had been no attacks. They saw no evidence of a perpetrator. No evidence of any kind of weapon. They couldn't find any of the things that police usually look for. And we, as scientists that were working in parallel . . . we reached the conclusion that the whole narrative I described at the beginning was impossible."

I can tell by his demeanor that it's important for him to be able to methodically go over all of the scientific reasons he found Havana Syndrome claims so outlandish.

"We examined, carefully, if any kind of energy weapon could have caused this. We contacted physicists. Biophysicists. We contacted people in France. We contacted people from different universities. And everybody agreed that microwaves were impossible. Sound was impossible. There was no known source. If you look at all the energy sources, sound so loud to harm somebody, you first would have to destroy all the ear to damage the brain. But everybody had their ears intact. Second, you [would have] to burn the skin to get to the brain. Third, ultrasound—you'd have to place the probe right next to the person. So we looked at each and every one of the sources, and it didn't fit into what was described. Diplomats were saying, 'I was in a room, and I was targeted specifically from outside the room.' But it's not possible with any energy source we know of.

"I'm willing to take the leap," he goes on, "if there's evidence of another source, to say, 'This needs an explanation.' But the question is, What needs explanation? Because if there's no brain damage, if there's no hearing loss, if there's no new syndrome, if there's no evidence of any perpetrator, then why do I have to search for a mysterious cause and make that leap, and make the assumption that there has to be a mysterious energy weapon?"

The bigger question, perhaps, is how and why such an unscientific theory was given so much credibility in the US media and among public officials.

Valdés-Sosa tells me that "the principal information" US authorities relied upon to substantiate Havana Syndrome came from reports published in the *Journal of the American Medical Association* (*JAMA*). The first one, in March 2018, was titled "Neurological Manifestations Among US Government Personnel Reporting Directional Audible and Sensory Phenomena in Havana, Cuba." According to Valdés-Sosa, it illogically conferred validity to Havana Syndrome before setting out to first verify whether it even existed. The unscientific starting point was to "describe the neurological manifestations" after exposure to "an unknown energy source associated with auditory and sensory phenomena," even though "an unknown energy source" hadn't been identified.

After examining some of the reported Havana Syndrome cases, the scientists in *JAMA* reported, "The unique circumstances of these patients and the consistency of the clinical manifestations raised concern for a novel mechanism of a possible acquired brain injury from a directional exposure of undetermined etiology." That's a stretch. The most scientists could have reasonably concluded at that stage was that they saw brain issues in people who said they hadn't previously suffered concussions or other trauma. To theorize about a "novel mechanism" to blame with zero evidence that any such thing exists, is hardly scientific. A weak follow-up in July 2019 attempted to continue driving the same theory, still with no potential energy source identified.

Consider the vastly different approach the public health establishment gives to theories it wishes to disprove rather than promote. When millions of adverse event reports were filed after Covid vaccination, the general approach was to insist none of them were related to vaccines. But with Havana Syndrome, a connection was assumed without any evidence whatsoever.

I look up the names of the scientists behind the *JAMA* papers. *What is their agenda?* I find they're primarily from the University of Pennsylvania Perelman School of Medicine—home of the infamous Dr. Paul Offit (more on him a bit later). Based on experience, we can assume the university and its scientists can be counted on to follow government and pharmaceutical narratives when called upon to do so. According to the medical school's website, it "is consistently among the nation's top recipients of funding from the National Institutes of Health, with $550 million awarded in the 2022 fiscal year." That kind of cash can buy a great deal of sway. Often, academic institutions become beholden to the wants and desires of the federal government/Big Pharma complex that gives them so much money.

Indeed, to hypothesize why Havana Syndrome studies with the unscientific approach ever got done, it makes sense to consider the timing and political climate. President Obama had previously eased restrictions on Cuba. Now anti-Cuba forces—including some inside the Trump administration—were trying to convince President Trump to clamp back down. Is it possible they used Havana Syndrome as a convenient and high-profile way to villainize Cuba and cast it in a negative light? A 2018 editorial in Cuba's Communist Party newspaper accused the US of as much. It called the Havana Syndrome hype part of a "permanent impeachment orchestrated by the US State Department."

"One would have to ask who is most affected by these arrows being launched against third countries without any solid evidence and the crude and meaningless way in which they blame and allude to Russia and China, after it is still not proved to have happened, even after months of rigorous investigation," the editorial says.

Whatever the case, Trump did remain tougher on Cuba than his predecessor and continued to tighten up on restrictions throughout his presidency.

Valdés-Sosa says the American scientific work in *JAMA* was "rejected" by "the international scientific community" because it was so fraught with issues. He and his team tried to contact the University of Pennsylvania authors to discuss what the Cubans saw as gaping flaws in the two studies. "We haven't gotten a response from the people officially involved in the study by the Pennsylvania team," Valdés-Sosa says. "We wrote to them, we never were able to talk to them."

Valdés-Sosa and his team at the Cuban Academy of Sciences came out with their own findings. "We published our report saying that we believe that the whole narrative was false. Then we added in the information received from the Cuban police, which said that the FBI and the Canadian police found no evidence of attacks or of perpetrators. They found no evidence of any foreign agency attacking US diplomats. And then you look at the scientific part, it's impossible. Then the conclusion reaches there's no real basis for this whole narrative of Havana Syndrome," he says.

Think what you will about Cuba's initial findings. After all, there's no reason to blindly accept their views more than anyone else's. But once the US and other examiners around the world ended up empty-handed on Havana Syndrome evidence in 2023, one might think the original *JAMA* papers that prematurely validated it would be retracted—or at least revised with the crucial new information. As of this writing, that hasn't happened.

CHAPTER 4

The Frog Professor

What happened to Tyrone Hayes shouldn't happen to anybody. Yet it's become a disturbingly common story in science—even if not one that's discussed much in the media.

Hayes was an assistant professor on the biology faculty at the University of California, Berkeley, in 1997, when he was hired to conduct research for pharmaceutical and chemical giant Novartis. The research involved a controversial herbicide called atrazine, routinely used on corn crops, and one of the most common contaminants in America's drinking water. Hayes believes his scientific findings subjected him to the kind of corporate spycraft and intimidation you used to only see in the movies or read about in novels.

For some important context, it helps to have an idea of how big and powerful businesses like Novartis are. In 2022, Novartis reported having $51.6 billion in revenue, with about half of that—$24.1 billion—in profits. In 2000, Novartis spun off the agricultural chemical part of the company handling atrazine into a newly created firm called Syngenta, which was sold to the communist Chinese government in 2017. Today Syngenta reports annual sales of more than $16 billion in pesticides, seeds, and other products.

■ ■ ■

Money = Influence

Like many pharmaceutical and chemical companies, Novartis and Syngenta hire researchers at hundreds of colleges and universities around the world. Critics say when companies fund academic research, it results in the "buying up" of academic institutions. Companies flood the schools with cash that's used to pay professors and researchers, upgrade labs, and help universities make a big name for themselves. The campus scientists are likely to avoid doing studies that could unearth safety risks with their sponsors' products. And when pharmaceutical firms give money to medical schools, the companies can influence what budding young doctors learn. Unsurprisingly, a heavy focus in med school today is put on prescribing medicine. Natural solutions and drug adverse events are glossed over and explained away.

Corporate funding is one of the two biggest factors involved in swaying today's scientific research. The other is government. Each year, the National Institutes of Health (NIH) doles out more than $33 billion in taxpayer-funded grants to over 58,000 entities. That purchases great control over America's research landscape. NIH tends to give the most money to studies that will benefit its pharmaceutical industry partners. NIH is seen as less likely to fund big studies that could unearth inconvenient facts or harm sales of favored medical products.

To put it succinctly, Adam Andrzejewski of the watchdog group Open the Books tells me that NIH "accumulates great influence with the power to decide which scientists and projects get all those taxpayer billions. . . . Buys you a lot of friends, buys you a lot of allies, and there's great incentive to stay on the establishment narratives that NIH disseminates on public health policy."

So what if a university scientist receiving government or corporate funds happens to stumble across something concerning? What if he "goes rogue" and follows the science down a path that may not be in the interest of a company's bottom line? *He must be taught a lesson.* Perhaps he'll be discredited and smeared. Monitored. Attacked academically by his peers. Maybe the institution where he works will fall under pressure to fire him or defund his research. There aren't many who can stand up to the kind of pressure that the government-pharmaceutical partnership can apply with help from the media. As I wrote in *The Smear*, there's an entire cottage industry made up of experts-for-hire, nonprofits, LLCs, super PACs, websites, foundations, PR companies, global law firms, and crisis management specialists who make their living destroying those who dare to come down on the wrong side.

That's what happened to Professor Hayes.

Feminized Frogs

In 2023, I head from the East Coast to California to meet Professor Hayes on Berkeley's campus. It's a chilly, wet day but our visit takes place indoors. We're seated in a frog-laden biology lab where he instructs me on how to not contaminate the active research specimens. Hayes is black, heavyset, and fashion-conscious. He's wearing a black suit, black shirt, red and black tie, with a coordinating red and black plaid shawl draped over his shoulders. A neat, tight haircut, fingernails cut short and painted red; an ornate, silver pendant earring with a red jewel hanging from his triple-pierced right ear. Tendrils of his Viking-style gray-tinged, braided beard drape down to his waist. He's at once both ready with a smile and guarded.

First, in our interview, he provides a definition of "developmental endocrinologist."

"That means I'm interested in the role of hormones in regulating development," he tells me. "Particularly, I work on amphibians. And my interests are in control and development of metamorphosis . . . the transition of a tadpole to a frog. But also during that transformation, I'm interested in how the decision to become male, or to develop testes, is made relative to the decision to develop ovaries."

When Hayes was hired to conduct research for Novartis in 1997, atrazine was the company's top-selling agrichemical. He says his mission was to figure out if atrazine somehow interfered with hormonal regulation and development.

"What did you learn early on?" I ask.

"Early on, we showed that atrazine demasculinized animals as well as feminized animals. So genetic males, which should develop testes and a certain type of larynx—voice box for attracting females—that aspect was inhibited in those genetic males; oftentimes they developed ovaries and actually transformed into females," he says.

Atrazine turns boy frogs into girls?

Not only that, but Hayes discovered that some atrazine-exposed frogs developed both male testes and female ovaries. Others had multiple, deformed testes.

The revelations about frogs was worrisome. But it's what the research implied about people that made Hayes and his information so—*dangerous.* After all, atrazine runs off crops and gets into our groundwater. It also can travel into the atmosphere on dust and end up in rainwater. Animals live in the contaminated water, and people drink it. The stakes were incredibly high. At the time of Hayes's studies for Novartis, the government had recently announced it was reviewing atrazine's safety, meaning it could be restricted or even banned.

"Was there resistance to making the findings that you had public at the time?" I ask.

"Yes. [Novartis] and the consulting firm that I was under contract for was not interested in me publishing the work," he says flatly. "And in fact, it was covered in my contract that I needed permission from the company to publish the work." But he balked at the idea of keeping his important findings hidden.

"It was against my training to not openly share and present data, or sort of hide and sequester data so it was not available. That's just not how I was trained as a scientist," he tells me. "I became uncomfortable with some of the things that the company asked me to do . . . and some of the, quite honestly, manipulations of the data that they were requesting."

So, in November 2000, Hayes quit working for Novartis. "I decided to repeat the work without their funding so that I had freedom to publish and to discuss those data with the scientists outside of my lab and outside of my university," he says. Around that time, Novartis spinoff Syngenta took over atrazine.

His additional studies on atrazine produced similar findings, not only in frogs but also in other species. The work was published in prestigious peer-reviewed journals. In the *Proceedings of the National Academy of Sciences*, Hayes reported that "hermaphroditism," the condition of having both female ovaries and male testes, occurred in frogs exposed to atrazine at levels thirty times *below* what the Environmental Protection Agency says is safe for drinking water. As his work was publicized, Hayes soon came to feel like Syngenta Public Enemy #1.

"The company tried very hard to get the work retracted or to publish counterwork or actually contact the university to try to get them to stop me from publishing," he says. "They did a lot of things to me. There was a pressure to inappropriately manipulate data to make it appear not as bad as it was. They hired other scientists to conduct experiments in ways that weren't informative. And even some of the studies that they conducted found the same results, but they wrote something different. They tried to get my work retracted from journals. They actually launched personal attacks on me. And they had plans, revealed through documents obtained in a legal hearing, to harass and pursue my students and my family."

Hayes's accusations might be written off as paranoia and exaggeration—certainly the company he accuses would likely prefer it that way—but for documents revealed years later. The documents came out as part of lawsuits filed by dozens of cities. The lawsuits accused the company of contaminating the cities' drinking water and "concealing atrazine's true dangerous nature." When the lawsuits settled in 2012, reams of internal Syngenta documents were unsealed. They exposed the alarming campaign Syngenta mounted against Hayes and his work. The *New Yorker* published a lengthy article about it written by Rachel Aviv, back when the mainstream press still investigated such things.

As part of her reporting, Aviv quoted from Syngenta documents. They provided chapter and verse on shocking tactics widely deployed by corporate and political interests against fact-finding journalists, advocates, and researchers. In this instance, the tactics included conducting opposition research to identify perceived weaknesses to discredit the target; following and verbally intimidating him; getting at him through his employer; threatening his livelihood; enlisting academics to speak or write publicly against their colleague, Hayes; manipulating Internet searches to smear him, make his work invisible, or point to opposing studies funded by the company; even investigating his wife.

One of the Syngenta documents was a spiral notebook. In it, Syngenta's PR team had drafted a list of ways to attack the uncooperative assistant professor. "[D]iscredit Hayes," reads one item. Syngenta communications manager Sherry Ford wrote that the company could "prevent citing of [Hayes's] data by revealing him as noncredible," "have his work audited by 3rd party," "ask journals to retract," and "set trap to entice him to sue." Ford also wrote about looking for ways to "exploit Hayes's faults/problems," and speculated that if he were "involved in scandal, enviros [environmentalists] will drop him."

In addition, Syngenta reportedly hired an outside contractor to do a "psychological profile" of Hayes. Ford wrote that Hayes "grew up in world (S.C.) that wouldn't accept him," "needs adulation," "doesn't sleep," was "scarred for life." "What's motivating Hayes?—basic question."

To aid in the efforts, the *New Yorker*'s Aviv reported, Syngenta's PR team identified one hundred "supportive third-party stakeholders" to enlist for help, including two dozen professors. Another idea floated was to control what Internet users would find if they looked online for details of what Hayes had unearthed. "[P]urchase 'Tyrone Hayes' as a search word on the internet, so that any time someone searches for Tyrone's material, the first thing they see is our material," wrote Syngenta's PR team, later proposing to also buy the search phrases "amphibian hayes," "atrazine frogs," and "frog feminization." "[I]nvestigate wife," wrote Syngenta's Ford.

Syngenta would later say that many of the documents "refer to ideas that were never implemented," according to Aviv in the *New Yorker*.

The fact remains that significant defenses of atrazine, and critiques of Hayes and his work, came from sources whose ties to Syngenta weren't always made obvious. Here are four examples.

The first comes through a "working paper" authored by Don Coursey, an economist at the Harris School of Public Policy at the University of Chicago. At a National Press Club event arranged in Washington, DC, Coursey announced that "a ban on atrazine at the national level will have a devastating, devastating effect upon the U.S. corn economy." It turns out Syngenta had hired Coursey at $500 an hour to embark upon an economic impact study, supplying him with selected data and research, and editing his work. (Coursey did disclose his funding source.)

The second example: Documents show that in October 2009, a PR group called White House Writers Group worked on behalf of Syngenta to write several op-eds. They were edited and approved by Syngenta's legal counsel, apparently intended to be published before an upcoming court date in a lawsuit against atrazine. All Syngenta needed were people it could convince to sign their names to the op-eds, as if they'd written them. "Future of Illinois on Trial in Madison Lawsuit" was one of the op-ed's proposed titles. The corporate authors used a style that made it sound like the op-ed was written by someone in the local community. "As Illinois struggles to rise from the worst downturn in 60 years," reads the curated but as-yet unsigned op-ed, "Texas trial lawyers and their St. Louis partner are about to deal us another body blow. If they succeed, farmers will be devastated, city jobs will be destroyed, and our regional tax base will be eroded." Syngenta's PR team ultimately got editorials published in the *Des Moines Register, Rochester Post-Bulletin, St. Cloud Times,* and *Washington Times* bylined by "third-party allies." The editorials defended atrazine and attacked critics.

A third example of how Syngenta orchestrated attacks against Hayes involves the Center for Regulatory Effectiveness. In 2002 the center petitioned the EPA to disregard Hayes and his atrazine research,

> stating, "Hayes has killed and continues to kill thousands of frogs in unvalidated tests that have no proven value." But the Center for Regulatory Effectiveness wasn't a neutral observer. It was run by a paid Syngenta consultant and congressional lobbyist, as Aviv reported in the *New Yorker.*

> A fourth example, and perhaps the most instructive one, is that of freelance columnist Steven Milloy. Milloy wrote an article entitled "Freaky-Frog Fraud," published online at Fox News in 2015. The point was to discredit Hayes's work. Milloy's writing style reads like a template for fake "fact-checks" similar to the ones later used to undercut safety concerns about Covid vaccines, most everything about Donald Trump, and a host of other topics. Milloy uses trademark terms that often provide a signal that propaganda or smear efforts are in use: "hijinks," "shoddy," "lame," "scary," and "fearmongering."

First Milloy refers to Hayes as a "junk scientist," falsely implying that his research is somehow discredited or compromised. Milloy then questions Hayes's credibility, dissecting the relevant frog study.

"We can't even be sure the concentrations of atrazine in the lab [frog water] tanks are what Hayes claims," writes Milloy. He then ascribes an unscientific motivation to Hayes, adding, "Hayes seems to be determined to scare the public about atrazine." This ignores the fact that Hayes had originally embarked upon the research at the behest of Novartis. Are we now to believe that Novartis had hired a bad scientist at a prestigious institution who, for reasons unknown, then developed an inexplicable desire to harm his sponsor and "scare the public"? And that his findings are only to be believed if they came down in favor of the herbicide?

Milloy deploys another common strategy frequently used by propagandists: attacking scientists or journalists as "anti-" something. Milloy writes, "Hayes' anti-atrazine effort smacks of the same sort of scientific hijinks committed by Tulane researchers in 1996—anti-chemical fearmongering eventually determined by federal officials to be scientific misconduct."

Finally, Milloy implies that because Hayes received research funds from environmental groups after he severed from Novartis, his findings are compromised. Again, this makes no sense. Little to no research is self-funded by scientists or paid for by disinterested parties. Milloy doesn't suggest that Novartis-funded research should be disbelieved: only studies supported by the other side. Even more relevant, while Milloy is attacking Hayes's funding sources, which Hayes disclosed, he's hiding his own. Milloy has been registered as a lobbyist, paid to influence decision-makers, though that fact wasn't always admitted in his writings. More to the point, Milloy had a long-standing financial relationship with Syngenta, which wasn't acknowledged in his Fox article.

Milloy's ties to Syngenta were laid out in the documents unsealed as a result of the atrazine litigation. On December 3, 2004, Milloy emailed Syngenta's Beth Carroll asking for money for his group, the Free Enterprise Education Institute.

"The Free Enterprise Education Institute requests a grant in the amount of $15,000 for its ongoing Atrazine stewardship cost-benefit analysis project," Milloy writes.

It's not a one-time request. On August 6, 2008, Milloy again emails Beth Carroll to ask Syngenta for cash.

"Hi Beth . . . is it too early to bother you for a grant?"

"No it's not too early for an invoice," Syngenta's Carroll replies an hour and a half later. "Send it for $25,000 and I'll see what I can do."

Milloy sends the invoice and writes, "Please note that the nonprofit we are working through now is called the 'National Center for Public Policy Research.' Send the check to me as usual and I'll take care of it. Steve."

Later that month, Syngenta's Carroll asks Milloy to remove the word "atrazine" from the title of his invoice and replace it with "pesticide stewardship cost benefit analysis project."

Through these tactics and more, Syngenta helped convince the EPA to kick the can down the road for years on whether to restrict atrazine, while the company continued to rake in billions. As time passed, more studies confirmed atrazine's harmful effects. But that research fell under a steady barrage of challenges from Syngenta.

In 2012, Syngenta settled the drinking water lawsuits and agreed to pay $105 million but denied any wrongdoing. The company continues to insist that atrazine does not harm people in normal, real-world exposures.

I ask Hayes to explain what the EPA finally decided about atrazine more than twenty years after his first study on its impact. He tells me that in 2020, "the EPA released its final assessment and concluded that atrazine is likely to affect 54 percent of all species and 42 percent of all critical habitats. In that same year, the EPA reregistered atrazine for use."

"Said it's okay to use?" I ask.

"Said it's okay to use," he answers. "Despite this preponderance of science showing that it has adverse effects on animals."

"To what do you attribute that?"

"Makes somebody a lot of money," he replies. "And there's a big lobby and industry that has a big influence over decisions like the ones that the EPA have to make."

Of course the purpose of this chapter isn't to definitively litigate atrazine's safety. The lesson is in how the industry was able to so confuse the scientific analysis, it becomes nearly impossible for a layperson to know what to believe. The lesson is how a corporate team mounted a sustained, multipronged personal attack against a scientist, his reputation, and his career. It's how the team conspired to use media, academics, social media, and the Internet to control the message.

From the view of Syngenta officials, they simply used their resources to set the record straight on a product they claim is safe, and to defend it from inaccurate and unfair attacks.

From Hayes's view, "We know that [atrazine] has environmental impacts, impacts on environmental health and public health, in every vertebrate animal that's ever been studied, unless the industry is throwing their money at the study. I'm completely confident in the science that atrazine has these effects in fish, amphibians, reptiles, birds, mammals, including humans."

As my time with the frog professor at Berkeley draws to a close, I can't resist addressing an elephant in the room.

"Have you wondered," I ask, "if [atrazine] is a hormone disruptor, if this can be playing any role in what we're seeing happening in our youth today? When there are a lot of boys who say they feel like girls and girls who say they feel like boys?"

"[It's] very likely that chemicals like atrazine that can influence your hormonal balance—and we know it does so in humans—that potentially could influence things like sex or gender identity and orientation," says Hayes.

Sounds like front-page news to me. But you won't see it on the front pages. Hayes then provides some important context that gives food for thought. "Atrazine is the poster child because we know what it does. We know it's not good and it's everywhere. [But] we have something like 80,000 human-made chemicals in the world. Most of them haven't been studied in the level of detail that atrazine has. If you go through the literature right now, you'd get glyphosate or Roundup, metolachlor, atrazine, maybe one or two other compounds. Most of the other compounds we use in agriculture, we know nothing about what they do. So yeah. We're missing a lot of information."

As for personal lessons? "You know, I have very few regrets in my life," Hayes tells me wistfully. "If I were doing it all over, so to speak, I'd be smarter about some things. There are some contracts I maybe wouldn't have signed. But at the same time, I learned from all those things, and you know, I ask my students all the time, 'If you get 100 percent on an exam, did you really learn anything?' So you have to get a few things wrong in order to challenge yourself and actually learn something."

The Trans Lobby

When a scientific issue goes from zero to 60 practically overnight—grabbing headlines, political real estate, and funding priorities—you can bet there's an invisible hand at work. The incredible transgender lobby proves the point.

In December 2021, Harry Potter author J. K. Rowling made news with a tweet. She criticized efforts to replace the word "women" with the trans-friendly phrase "people who menstruate." Rowling was called "transphobic" and even threatened with violence, she says.

Hers is hardly a fringe view. Recent surveys by Rasmussen Reports and Public Religion Research Institute indicate a majority of Americans, at least two-thirds, say there are only two genders: male and female. But expressing that majority view has become increasingly forbidden. So I set out to learn who is pulling strings behind the scenes.

First, I speak with Gregory Angelo, a gay man and president of the New Tolerance campaign, a group fighting what he calls intolerance double standards.

"How such a small slice of America, or what was a very small slice of America, could take front and center stage in education and media and politics over a pretty short period of time is an interesting question," I posit.

"Yeah, it's curious," Angelo agrees. "People who think that this is a grassroots movement that is giving rise to the transgender culture in the United States, this transgender moment that people say that we're having, do not understand that this is more of a top-down dynamic that is at play, where you have really just a handful of organizations and [lesbian gay bisexual trans] advocacy organizations that are driving the agenda. But because they have such sway and such money, they're influencing everything from schools to politicians, to corporate America."

Who, exactly, has been bankrolling the transgender agenda as it's been catapulted to the top of our international priorities—and why? By now you may be able to guess. It's some of the same big-money interests that are guiding the public narrative on many other fronts.

One example can be found in Lupron. It's a prescription drug used to suppress puberty in some "transgender" children who wish to be the opposite sex, though it isn't FDA-approved for that. One Lupron injection for children reportedly costs as much as $10,000. Some treatment plans call for monthly injections for years. AbbVie, the company that makes money selling Lupron, happens to generously support the transgender movement.

In 2020, AbbVie gave "Silver Tier" support of $50,000–$100,000 to The Trevor Project, a controversial gay, bisexual, and transgender advocacy group. After receiving the money, The Trevor Project then produced an important study that seemed to make the case that children should have access to "gender-affirming care," which would include Lupron. The study was even cited by the Biden administration.

There's an even bigger medical money interest surrounding the trans movement that I haven't seen discussed very much. It involves a sad reality: the transgender HIV epidemic. A CDC survey of seven major US cities—Atlanta, Los Angeles, New Orleans, New York City, Philadelphia, San Francisco, and Seattle—finds that 42.2 percent of transgender people who are men living life as women are infected with HIV, the virus that causes AIDS. The rate is even higher among transgender blacks: *most* of them—61.9 percent—are HIV-positive. And the HIV rates are even higher among transgender prostitutes. That adds up to a fantastically lucrative market for the makers of HIV medicine, like Gilead Sciences. Gilead has developed eleven HIV medications now on the market earning over $1.5 billion a month.

Pulling the thread further, I learn that Gilead happens to be the single biggest known funder of trans activist groups, providing $6.1 million in 2017 and 2018. Like AbbVie, Gilead is also listed as a "Silver Tier" supporter of The Trevor Project in 2020. And on the project's website in February 2024, Gilead is listed as a $250,000–$500,000 supporter. Gilead says its goal is to "support communities that are disproportionally impacted by diseases aligned with our therapeutic areas of focus."

CHAPTER 5

Ten More Eye-Openers

Now that you've had the shot, here's the chaser. I've carefully selected some seminal moments along my path to enlightenment into the many ways that medical information is manipulated. Think of this as a CliffsNotes version of how I've managed to break international news on health topics seemingly overlooked by the presumed specialists in medical, health, and science beat reporting. These examples help explain how I became equipped to distinguish between facts and propaganda in science. They can help you too.

1. Viagra and Blindness

On April 1, 2005, a brief report crosses the news wires feeding into the CBS Washington, DC, newsroom. I scan it and then reread it with more interest. It tells of some research by an ophthalmologist named Dr. Howard Pomeranz.

Dr. Pomeranz, it seems, has identified a cluster of cases of a mysterious and previously rare form of blindness called "nonarteritic anterior ischemic optic neuropathy," or NAION, among some of his male patients. On his own, Dr. Pomeranz decided to launch a scientific inquiry and discovered that the afflicted patients had one thing in common: they were all taking the erectile dysfunction drug Viagra, and their blindness episodes occurred within thirty-six hours of their taking a dose.

Though the item isn't given much prominence on the news wires, it strikes me as *hugely* significant. Viagra is a blockbuster drug taken by untold millions. I point out the article to some of my newsroom colleagues. *Maybe we should flag it to the New York editorial staff in case they want to report it on the* CBS Evening News *or look into it further,* I say. They laugh and ask if the story is an April Fools' joke, seeing as it happens to be April first. *You know, Viagra . . . the myth about masturbation causing blindness . . . blindness being reported in these men . . .* Nobody but me seems curious beyond the first laugh.

I start digging around. I reach out to a law firm that has a strong record of success in exposing prescription drug side effects. I contact Dr. Pomeranz. And I search the government database that collects reports of drug adverse events.

I'm stunned by what I find: an unambiguous pattern of blindness reported among patients taking Viagra. I also notice a clear trend of deafness reported in Viagra patients. But, for the moment, I remain focused on the symptom that has supporting data published by Dr. Pomeranz—the vision problems. Equally as important is a larger question: How have the experts at the FDA, who are tasked with reviewing the data and monitoring drugs for safety, missed the obvious signals? Have they not read the disturbing data? Furthermore, drug companies have full access to FDA data as well as their own adverse event reports from consumers. So why hasn't Pfizer already warned the public? For that matter, why did it take Dr. Pomeranz, an ophthalmologist, to make the connection between Viagra and blindness, rather than the prescribing urologists who should have noticed first?

Over the next few days, I call the FDA to ask for more information and an on-camera interview. The agency doesn't want to do an interview, but the press shop agrees to connect me with the FDA officials tackling Viagra safety. In a phone call, the officials surprise me by confessing they're in the process of negotiating with Pfizer to add a warning about the risk of blindness to Viagra's label. Suddenly, my story becomes bigger. The FDA isn't going to try to deny that Viagra can blind men. They're acknowledging it!

I'm ready to break the news. I put the finishing touches on my story, have it legally reviewed by my CBS attorneys, and call the *Evening News* executive producer in New York to tell him it's ready to go. The following evening, I report the story on the *CBS Evening News* and it's picked up globally. Blindness warnings are soon officially added to Viagra. Two years later, warnings for deafness are added as well.

As far back as 2000, Dr. Pomeranz had been publishing about the link between Viagra and blindness. How many men had needlessly lost their eyesight between 2000 and the time the warning was finally issued in 2005 because experts in government and the press had failed at their jobs? Even with the warnings, the media hasn't done a very good job reporting out the facts. In 2022, a large study of 200,000 men found that erectile disfunction drugs like Viagra, Cialis, Levitra, and Stendra can more than double the risk of conditions that can lead to blindness. Most men don't know this. And from what I gather, many prescribing doctors aren't warning them about it or watching for it.

Lessons Learned: Don't assume the experts will be first to alert the public about drug safety issues. The FDA isn't necessarily monitoring or acting promptly on adverse event reports. Likewise with drugmakers, and even the prescribing doctors. There's every chance that a curious reporter or physician who is outside the specialty will unearth crucial information before the inside experts do.

2. Ghostwriting

It's an industry insider who first clued me in on a shocking practice that I'll bet most patients, and even most doctors, don't know about. One medical writer refers to it as "the dirty little secret of medical publishing."

Ghostwriting.

Ghostwriting is when a doctor who appears to be independent signs his name to a medical journal article promoting a drug or treatment. The article is actually written by the drugmaker or its agents.

The signing doctor is more or less a hired hand who's paid for his signature. Those who read the article are none the wiser about the true nature of the content because the source isn't disclosed.

When I learned how widespread medical ghostwriting is, I found it hard to believe it's legal. At the very least, it's ethically troubling in a profession that should be all about ethics. There's no doubt the practice can be misleading. There are numerous well-documented instances in which companies used ghostwritten articles to market dangerous drugs. The articles exaggerated benefits and downplayed risks. The fact that ghostwriting *is* legal can be explained by the fact that those benefiting are often the very same entities that write the rules saying it's perfectly okay. And they often donate to political decision-makers to get them to say it's okay too.

When Wyeth-Ayerst wanted to create demand for its "fen-phen" diet drug Redux, it hired a middleman named Excerpta Medica. Excerpta Medica agreed to get nine articles promoting Redux published in medical journals. Excerpta wrote the articles, hired doctors to review and sign them, and then submitted them for publication—with no mention that Wyeth was bankrolling the whole thing. Basically, the articles were paid ads for Redux disguised as scholarly work.

During a subsequent lawsuit over Redux injuries, former Wyeth executive Jo Alene Dolan defended the ghostwriting. She claimed all drug companies do it, and that it doesn't mean the articles aren't accurate. For its part, Wyeth's middleman, Excerpta Medica, claims it doesn't ghostwrite; it "facilitates." Ultimately, Redux was pulled from the market due to heart and lung problems.

Wyeth also paid ghostwriters for more than forty articles between 1998 and 2005 promoting its hormone replacement drugs Premarin and Prempro. Use of those medicines was later greatly limited for safety reasons. Women taking them had a higher risk of heart disease, stroke, breast cancer, and dementia. In subsequent lawsuits, injured patients claimed the ghostwritten articles emphasized the drugs' supposed benefits and downplayed risks while generating billions of dollars in annual revenue for Wyeth.

In a third high-profile example from the same time period, drug company Merck used ghostwriters to market the painkiller Vioxx and quickly earned billions. Vioxx was later pulled from the market due to increased risk of heart attacks and strokes.

As you'll learn in a moment, ghostwriting is far from the only weakness to be found within the pages of our revered medical journals.

Lessons Learned: Scientific literature in peer-reviewed, published medical journals may be conflicted. Some doctors are for sale.

Which leads us to the next example. A close cousin to ghostwriting is the secretive use of paid medical experts.

3. Hired Guns

Early on in my reporting on pharmaceutical controversies, I noticed that drug companies never agreed to give me on-camera interviews. We'd go through a routine. They'd typically refer me to speak to a specific "independent doctor" who they seemed oddly confident would defend their drug. From the company's perspective, it generated far more trust to have an "independent doctor" dispel concerns about a medicine than to have the company defend its own product.

Considering that I was reporting on drugs that had obvious safety questions, I came to wonder why any truly independent physician would go out on a limb and blindly back the medicine in question. And how did the drug companies know these particular doctors would happily speak to the media? One day, I got the idea to do an Internet search on one of the doctors. *Bingo!* I discovered he was a paid consultant for the drugmaker—but neither he nor the company had revealed that relationship!

A conflict of interest exists when a doctor has financial ties to a drug he's testing, prescribing, or offering opinions on. When a doctor stands to gain or lose based on a drug's success, he has motivation to promote the drug. That relationship can inject bias even if the doctor isn't consciously skewing his opinions.

What I'd discovered was a classic, undisclosed conflict of interest.

How do the drug companies pull it off? And why do the doctors agree? There are numerous tactics used to cultivate friendly and mutually beneficial relationships. The companies pay doctors to travel to exotic locations and give presentations at professional conferences. They pay doctors handsomely as consultants or speakers. They hire researchers and their academic institutions to lead scientific studies. And, as we've learned, they hire doctors as ghostwriters. Taking money from a drug company doesn't automatically negate a physician's opinion given for a news story. But news reporters should disclose the financial relationship so that viewers can appropriately weigh the doctor's views. Too often, journalists don't even bother to ask the question: *Does the doctor they're interviewing have financial ties to the drug he's defending?*

I've conducted hundreds of Internet searches over the years uncovering financial ties between doctors and drug companies that seem to never be disclosed in media stories. I came to learn that if I simply ask the physician directly, he usually acknowledges a financial relationship. I came to wonder why so many medical beat reporters would quote the same doctors as I did without asking the basic question or, if the reporters were aware of the conflict of interest, why they chose not to disclose it to readers and viewers.

In 2004, the watchdog group Public Citizen is pushing to get the cholesterol-lowering statin drug Crestor removed from the market due to kidney failure in some patients. For a story for the *CBS Evening News*, I ask Crestor's maker, AstraZeneca, for an on-camera interview. The company declines. However, I start getting unsolicited calls from doctors who say *they* would be happy to give interviews defending Crestor. In an unsolicited email, one of the doctors urges me to interview him to "ensure ALL the information about this important class of medication gets out to the public, and not just a selective interpretation of data." I also find myself getting lobbied by a Cleveland Clinic Foundation nephrologist who insists, via another unsolicited email, that some safety concerns about Crestor are false and that it's "imperative" that my story be "both factual and accurate." Next comes a bunch of local newspaper clippings sent to me by AstraZeneca. In the clipped articles, doctors around the country are saying positive things about Crestor. I conduct my conflict of interest search online, cross-referencing the doctors' names with AstraZeneca. I quickly find that each and every one of them has a financial relationship with AstraZeneca, though it hadn't been disclosed in any of the articles. Not one.

I contact AstraZeneca's press folks and let them know I've figured out what's going on. I say that I'd be happy to interview one of their recommended doctors on camera, but that I will mention in my story that the doctor is a spokesman representing AstraZeneca's views. As an investigative reporter, not a medical reporter, I'm mystified that I'm apparently the first journalist to take this tack. An AstraZeneca PR representative sputters on the phone when I tell her. She doesn't seem to want me to interview one of their recommended doctors on camera if I'm going to disclose to viewers that he works for AstraZeneca. Clearly, the company is concerned about the public knowing the so-called "independent experts" aren't really independent. The company is fighting for the life of what it hopes will become a blockbuster drug. Billions of dollars in sales are at stake.

As my story research continues, AstraZeneca continues to bombard me with additional referrals to people they say I should speak with off camera. They won't do TV interviews but want to spin me behind the scenes. I'm fielding near-constant calls and pressure from the company, a PR firm working for it, its physician "experts," and a big-money crisis management firm hired to spin away negative news on Crestor. When AstraZeneca feels like it's not making headway with me, their representatives go over my head and contact the executive producer of the *CBS Evening News* and the president of the news division to try to influence my reporting. I would learn years later that the CBS advertising division contacted the *CBS Evening News* executive producer on behalf of AstraZeneca to warn about my coverage exposing the dangers of Crestor and other statin drugs that had aired on the news. The executive producer told me that "one of the [CBS] sales bosses threatened" him "pretty harshly" and left him a "loud, angry voicemail saying '[Sharyl's] stories could really harm [CBS's ad] business.'" Back then, as now, CBS and the other news networks received tens of millions of dollars a year in ad money from the pharmaceutical industry. A news division's editorial decisions should be strictly firewalled from the corporate advertising department. In reality, they're not.

Later, an important scientific alert is issued about a possible dangerous side effect concerning the rotavirus vaccine for diarrhea in children, RotaTeq. I notify the editorial managers with the *CBS Evening News* in New York (I'm still working from the network's Washington, DC, bureau). The alert warns that scientists have found an increased risk of intussusception, or twisted bowel, a potentially fatal disorder, in children vaccinated with RotaTeq. There was a time when that news item would have sailed on the air without a second thought. After all, it was related to a drug commonly given in three doses to American children before they reach the age of eight months—for a disorder US children don't typically die from. The risks of vaccinating should be well examined and disclosed so that parents can make informed decisions. Shortly after I alert the CBS managers, one of them sends around a group email with our medical correspondent weighing in. It's important for you to know what I observed about medical correspondents at the networks. It seems to me they came to typically be used to tamp down news that can damage big pharmaceutical interests. To this day, I don't know if the medical correspondents do it out of ignorance, simply taking the government and corporate line on controversies because they're taught to do so by their pharmaceutical industry–designed medical curriculums and industry-enriched medical associations, or because they're actively serving a financial interest. Many appear to be reflexively accepting and unquestioning of "science" as presented to them almost as if it's their patriotic duty to be that way. They seem to assume every new thing promoted by establishment medicine equates to progress with no downside. In any event, in the group email, the CBS medical correspondent says that we shouldn't report the news about RotaTeq's potential risks. It seems the correspondent has consulted the famous Dr. Paul Offit of Children's Hospital of Philadelphia, and Dr. Offit assures him there's no cause for concern over RotaTeq.

I'm speechless. Nobody on the email thread, including the medical correspondent himself, seems to be aware that Dr. Offit isn't an independent authority on this subject—he's the inventor of the RotaTeq vaccine that's at issue! I reply to the email thread disclosing this crucial fact. I'm hoping the important news about RotaTeq will make it on the air. Parents should have the information. But a few minutes later, another email comes and it's more outrageous than the first. The CBS medical correspondent says, okay, yes, he called Dr. Offit back, and yes, yes, it turns out to be true—Dr. Offit is indeed the inventor of the vaccine that he's defending. But the medical correspondent adds that Dr. Offit "assured" him in this follow-up phone call that nothing worrisome about intussusception showed up in the original RotaTeq clinical studies. Therefore, says the medical correspondent, we need not report the news.

I have to restrain myself from "replying to all" something like, *Are you all stupid?* Instead, I observe, in my email reply that while it's fine for Dr. Offit to defend his medicine, we shouldn't censor the story entirely from our viewers. We should report on the safety alert and can certainly include that Dr. Offit, the vaccine's inventor, says there's no cause for concern.

I lose that battle. The RotaTeq story isn't told on the *CBS Evening News* that night. This is the beginning of a trend that's since become a widespread practice. Instead of representing various sides of a story, news managers decide which viewpoint is "correct" or valid on the front end (even when they aren't qualified to know), and they either represent only that one side, or declare that side to be correct and the other wrong. Or they censor the story entirely so that you can't have the information to make up your own mind.

Some years after Dr. Offit, public health officials, and other pharmaceutical interests "debunked" reports of intussusception risk with RotaTeq vaccine, the evidence got to be so overwhelming, intussusception warnings were finally added to the vaccine's label. I wondered how many children were harmed or had their vaccine injuries overlooked because the risk was denied for so long, and the media chose not to report it.

Lessons Learned: "Independent" doctors and researchers who defend or promote drugs are often working for the drugmaker without affirmatively disclosing their financial ties. Pharmaceutical companies spend unimaginable fortunes on public relations efforts to spin and influence news reporters, their bosses, and corporate news headquarters. News editors may block stories that are contrary to certain interests.

4. Hide and Seek

While my education into pharmaceutical industry strategies was expanding, I communicated with FDA scientists who had roles in approving applications for new drugs. One of those scientists told me he likened his job to that of a detective. That took me aback.

"The drug companies are always finding new ways to hide adverse events that happen in their clinical studies," he explained. "So my job is to figure out what they're covering up in their FDA applications." He said it was widely known inside the FDA that drug companies actively conspire to hide the dangers of their drugs from FDA reviewers. He said it was also widely known that such subterfuge by the companies, when discovered, was never punished.

By way of background, it's important to know that when a side effect emerges during a study, it's an extremely significant event. Most side effects don't come to light during clinical trials because the samples are too small, the studies are too short, and the test subjects are usually healthier, overall, than those who will actually be taking the drug once it's approved. It's typically only when thousands or millions of people use a medicine for years that distinct patterns emerge and serious adverse events are detected.

So when a side effect *does* turn up early in a small study, it should be treated as a major red flag. It could prove to be dangerously common in the real world. The very point of testing drugs in people is to figure out if that's the case, not assume the illness is unrelated to the medicine being studied.

When a company is trying to get FDA approval for its new drug, it has to be able to explain away any serious adverse events that do occur in its clinical trials. FDA scientists have told me they've caught drug companies using various tricks to do that. One is to bury them. That happened with the diabetes drug Rezulin. Rezulin was taken off the market in March 2000 due to reports of liver failure. Its maker, Warner-Lambert, ultimately credited (or blamed) my CBS investigations into Rezulin with getting the drug removed.

After the drug was withdrawn, one of the FDA scientists who worked on Rezulin's original approval told me he began racking his brain trying to figure out how he'd missed its potential to cause liver problems. He combed back through reams of material Warner-Lambert had originally submitted to the FDA. It turns out the liver signals were there all along, he told me, but cleverly hidden. How? Instead of grouping them under one clearly identifiable heading, where they would have added up to an obvious red flag, they were divided up into smaller, less alarming subcategories. For example, some clinical trial patients who suffered liver issues may have been listed under "hepatic" events, others under the category of "cirrhosis"-related. Still more could be grouped under the title of "jaundice." When split into these smaller classifications, no single category rose to the level of a worrisome liver signal. Had they been properly clustered together, it would have amounted to a five-alarm fire, according to the FDA official.

Another trick companies use to address adverse events in their clinical trials is to make them disappear altogether. *Poof!* One way they do that is by blaming a patient's adverse event on his preexisting factors. For example, a test subject may develop heart inflammation during a study. The drug company will say there's no reason to think their drug caused it. "The patient's grandfather had heart inflammation, so it's clearly a hereditary problem unrelated to the medicine," the company may say. The fact is, people with predispositions are often more likely to get hit by a medicine's adverse events. As with the smallpox vaccine, existing illnesses and genetic vulnerabilities can magnify the risk of side effects. But the drug company writes it off.

You should also know how common it is for scientists to manipulate data on the front end to reach a desired answer rather than genuinely seek the truth. One way this happens is though unscientific data dredging. That's when researchers collect a lot of data and keep analyzing it in different ways until they find the conclusion they wish to reach. Then they publish only the cherry-picked results. For example, a study may show a medicine doesn't work and could cause harmful side effects. But if the scientists keep changing the way they analyze the data, they may find a nugget that seems marketable. Maybe there's a short-term benefit for a subpopulation, such as young men. The researchers then publish only that positive result without disclosing the larger and more important findings.

Another tactic is for researchers to keep doing similar studies until they stumble upon one with good results. Maybe nine out of ten studies show no benefit of a drug, or even harm. But the only one that gets published is the one that happened to turn out positively.

Likewise, if a researcher is part of an effort to undermine a competing medicine or treatment that stands to undercut profits for their own company, they can design studies and crunch data in a way that's sure to produce negative results about the competition. Today there's no way for consumers to know how many studies are really done on a pill or injection before the one that actually gets published.

Of course, the obvious question is, Why does the FDA allow this? Why isn't there the harshest penalty for companies that fail to adequately disclose pre-market health issues? These are matters of life and death and should be treated accordingly. In fact, government officials have gotten caught actively cooperating with drug companies in a way that raises ethical questions. According to one scholarly analysis of the Rezulin disaster, "company and government documents . . . showed that officials from Warner-Lambert had collaborated closely with certain senior officials in the US Food and Drug Administration (FDA) during the approval process and later, when the company was being pressured to take the drug off the market." David Willman of the *Los Angeles Times,* who won a Pulitzer Prize for his reporting on Rezulin, elaborated: "senior government officials repeatedly played down the drug's propensity to cause liver failure and death. Before it was withdrawn on March 21, the FDA assured doctors and patients that Rezulin's potential benefits in lowering blood-sugar levels outweighed its grave risks. . . . As deputy director of the FDA's drug-evaluation center, [Dr. Murray M. 'Mac'] Lumpkin helped make Rezulin the nation's fastest-approved diabetes pill and, to the end, resisted its withdrawal."

Lessons Learned: Drug companies are not necessarily forthright in disclosing what they discover about the harm their medicines may cause. The studies they present may be one-sided and minimize risks. The FDA cannot be relied upon to promptly detect or reveal safety issues.

5. Controlling Data

I used to believe that a study in a peer-reviewed medical journal was as good as it gets when it comes to proving a drug works and is safe. Especially if it's a "randomized controlled trial." That's what establishment medicine tells us. You'll hear health officials parrot the phrase "gold standard" as in, *When a randomized controlled study is published in a peer-reviewed journal, that's the gold standard.* But I came to learn there are countless reasons why we cannot blindly accept that as definitive.

Researchers used to operate in a more scientific environment. Decades ago, valid study results were published even when they were negative. A lot can be learned from studies when a drug doesn't work or causes serious health problems. The information revealed can stop other researchers from wasting time or endangering study patients by following a similar path. But today, drug companies control the terms under which researchers can publish. If a university study reveals a medicine causes harm, the drugmaker that's paying the bills can block the study from being published. The company grants itself that power in the research contract. Additionally, drug companies have figured out how to stovepipe study data so that no single researcher has the complete picture. If research looks bad for a medicine and the drug company wishes to prevent the researchers from publishing it, it's easy to do so because no researcher has access to the full data he would need to disregard the company and publish the findings.

I covered a fascinating story along those lines. It started way back in May 1999 and involved a university researcher who was sued for millions of dollars after he published negative findings about a potential AIDS vaccine. The saga began when a big study on the vaccine was halted midstream because early results showed it just didn't work.

Under ethical guidelines, continuing the research would have been unethical because it could string along test patients with a pointless medicine and possible side effects. The lead authors on the halted study were Dr. James Kahn at the University of California, San Francisco, and Dr. Stephen Lagakos from Harvard's School of Public Health. They said they felt it was urgent for them to publish results on the failed AIDS vaccine so that other researchers could understand why it didn't work. It would keep other companies from wasting time and money on similar endeavors when more promising research could be pursued.

Trouble came when the maker of the AIDS vaccine, Immune Response, objected to Dr. Kahn publishing as he saw fit. Immune Response refused to provide him with the full data that would typically be needed for a study to get printed in a peer-reviewed journal. At the time, Dr. Kahn said, "We were really flabbergasted by this kind of response, and I've never seen anything like this."

It's unusual for scientists to go up against the companies paying them—even tougher today with tightened research contracts. However, Dr. Kahn decided to take on Immune Response. He published the AIDS vaccine study results in the prestigious *Journal of the American Medical Association*. The *JAMA* editors said they thought Immune Response was so out of line, they agreed to publish even without complete data.

In response to the publication, Immune Response sued Dr. Kahn and his university for $7 million. Imagine the chilling message something like that sends and how it can destroy the researcher who's targeted. There are lawyers and other extraordinary expenses to pay. There's stress at work and at home. Your university thinks you're a whole lot of trouble. Your bosses aren't happy to be at odds with drug companies providing lifeblood in terms of research funding.

At the time, the medical journals collectively stood up for Dr. Kahn. I don't think that would happen today. Back then, the world's top journals published a joint editorial warning that pharmaceutical companies exert too much influence over drug studies. "That line between the author's independent conclusions and the company's conclusions has been blurred," wrote Dr. Jeffrey Drazen of the *New England Journal of Medicine*. The editorial explained how drug companies had grown to virtually dictate terms of studies "that are not always in the best interests" of patients and science—and that "the results may be buried rather than published if they are unfavorable." More on that in a moment.

Immune Response dropped its $7 million case against Dr. Kahn the day the company was embarrassed by the joint medical journal editorial in January 2002. Unfortunately, drug company control over research has only become more heavy-handed since then.

A final note under this category is that drug companies and their researchers commonly tweak or manipulate the datasets, study parameters, or math to make negative results appear more positive. Though it's unethical and unscientific to change a study design midstream, and it invalidates the study for obvious reasons—it happens all the time. Medical journals, peer reviewers, and oversight bodies look the other way.

Lessons Learned: Drug companies retain tight controls over studies in such a way that may hide adverse events or exaggerate benefits. It cannot be assumed that drugs approved for market are as safe or effective as advertised.

6. Compromised Medical Journals

This next category of revelations along my educational journey is so important, it amounts to an international scandal and merits daily headlines. But you likely hear little to nothing about it in the news. It's the fact that the most well-respected medical journals routinely publish slanted, manipulated, misleading, and untrue information. That's not just my conclusion; it's something confirmed by independent published research and even by the leaders of the actual journals. For example, a research paper in 2023 found that about one-third of studies in neuroscience journals, and one-fourth of studies in medical journals, are "made up or plagiarized."

Harvard's Dr. Marcia Angell was one of the first to step forward publicly and blow the whistle on this dark secret. "I think physicians and the public have come to believe that drugs are much better and much safer than they really are," she tells me in an interview. Her statement is all the more remarkable because she's speaking out against the very journal she once led, the *New England Journal of Medicine* (*NEJM*). Dr. Angell worked at *NEJM* for twenty years and was its first female editor in chief, from 1999 to 2000.

"I came to the *New England Journal of Medicine* in 1979," Dr. Angell tells me. "Starting about then was when you saw the drug companies assert more and more control, until finally over the next couple of decades, they began to treat the researchers as hired hands. They would design the research themselves. You know, you can do a lot of mischief in how you design a trial. Or 'We'll test this drug, and we'll tell you whether it can be published or not.' And so if it's a positive study, it's published. If it's a negative study, it'll never see the light of day."

I say to Dr. Angell, "Most people probably think if an article is in a journal, it's probably written at a university and based on independent study, and that's that."

"It used to be that way, as you describe it. Pretty simple," she agrees. "And it began to change as the pharmaceutical industry became richer, more powerful, more influential, and began to take over the sponsorship of probably most clinical research."

Dr. Angell says that as editor in chief of *NEJM*, she tried to apply due diligence to the many studies that crossed her desk. She wanted the journal to maintain its credibility and properly police for accuracy and conflicts of interest. But she says it was a losing fight.

Dr. Angell tells me, "I would call up [a research author prior to publishing a study] and say, 'Okay, you've shown that your drug is pretty good. But [you claim] there's not a single side effect. [Yet] any drug that does anything is going to have some side effects.' And I had people say, 'Well, the [drug company] sponsor won't let me [publish that].' And so, I came to be extremely distrustful of most of the research that was published. We did our very best. We often rejected things because it was clearly biased. But anything we rejected always ended up in another journal."

Think about the implication. Your doctors and mine rely on research in medical journals as if it's the word of God. But it's often little more than paid advertising.

After Dr. Angell left the *New England Journal of Medicine* in 2000, she kept her eye on the journal landscape. She says the industry, largely co-opted by pharmaceutical interests, rejected meaningful reforms. In 2009, she wrote an article that famously declared, "It is simply no longer possible to believe much of the clinical research that is published." The editor in chief of the British journal *Lancet*, Dr. Richard Horton, has said much the same. In 2015, he wrote a scathing editorial declaring: "Much of the scientific literature, perhaps half, may simply be untrue; science has taken a turn towards darkness."

Half of the scientific literature may be untrue?

Years after my original interview with Dr. Angell, I return to her and ask her to describe the current state of the medical journal landscape.

"I think that that role[s] that the *New England Journal [of Medicine]* used to fill—one was the role of being skeptical, the other was the role of caring about the ethics of the whole system—I think the journal has given that up," she tells me. "The *New England Journal of Medicine* has given that up." For its part, the journal says experts put hundreds of hours of work in on each published paper to ensure they meet exacting standards.

A more recent assessment comes from journal editor and neuropsychologist Bernhard Sabel at the University of Magdeburg in Germany. In 2023, he authored an eye-opening study concluding that in 2020, "28% of all biomedical publications" were fake. According to Sabel, "as if you're buying a T-shirt in the shop, you can buy a paper for it to be published in the scientific journal." In an interview with NPR, he adds, "You can now go online, and you can see a title advertised, sign up here. Pay this and that much for it. There are papers that have fake photos. They have fake text. I presume many are automatically produced by artificial intelligence. And there are agencies who are specializing in this business, which creates a lot of junk in the scientific literature at a scope that is just unbelievable."

By 2015, I'd learned through experience that oddly glowing medical articles about controversial prescription medicines are often influenced if not wholly produced by the drugmakers.

When one admiring article about human papillomavirus (HPV)

cervical cancer vaccines is published in the British journal *Lancet*, I decide to do a little digging. The article is written by Dr. Sharon J. B. Hanley. It was published after the Japanese government decided to stop promoting HPV vaccines due to concerns about vaccine injuries. Dr. Hanley defends HPV vaccines in her article and criticizes the Japanese government. She also implies patients are incorrectly blaming HPV vaccines for unrelated illnesses.

What do I find after I poke around a little? Naturally, Dr. Hanley has financial ties to HPV vaccine makers, though she didn't disclose that in her medical journal article defending them. I discover she'd received funding from middlemen. They're entities supported by makers of both HPV vaccines: Gardasil and Cervarix. I also find that she'd previously admitted, elsewhere, to having received "lecture fees" from vaccine makers. But she discloses none of this in her *Lancet* article.

After I make these discoveries, I write *Lancet* to point out Dr. Hanley's undisclosed financial conflicts of interest. To its credit, the journal issues a correction based on my letter. Of course, it probably doesn't make much practical difference. The physicians who read the original article aren't going to comb back through the journal looking for corrections. I wonder, *Why don't journal editors or reviewers do at least a tiny bit of research, like I did, to uncover lapses in studies or conflicts of interest before the studies are published? Why don't journals ban researchers who violate the rules? Why don't journals retract the articles of the offending scientists?* That would go a long way toward stopping the shenanigans.

With all of this in mind, you may understand why I say this: When a journal study is retracted—meaning the journal more or less unpublishes it due to concerns raised about accuracy—it takes some deep analysis to understand what's really going on. It could be that, yes, the study is in fact bad. But it could also mean something else. It could mean that industry interests have successfully lobbied to have the study removed because the findings were harmful to their bottom line. A retraction could mean that an influential company lobbied to remove a good study that reflects well on a competitor. For these reasons, on occasion, retracted studies carry more truth than nonretracted studies on the same topic. In other words, a retraction can say more about who likes or dislikes the study's findings than the accuracy of the study.

A punctuation point to our lesson about the spotty record of revered medical journals comes from a story I reported on a few years back about the "Chocolate Diet." The Chocolate Diet was based on a study about the supposed benefits of eating chocolate. *Who wouldn't love a diet like that?* It was publicized with a slick marketing campaign and a catchy tune on the web, as a singer crooned the lyrics, "A scientist in Germanyyyyyyyy recommends choc-o-late treats . . ." But it turns out the whole Chocolate Diet was a hoax perpetrated by a journalist named John Bohannon to show how easy it is to get shoddy research published. When the scheme was exposed, Bohannon told reporters, "I already knew that there were fake journals who would publish this stuff, but [I wondered], would journalists pick it up and turn it into a big story? And the sad answer is yes—[a] *very* big [story]." The Chocolate Diet was promoted by mainstream outlets all over the world.

Lessons Learned: Information published in the most prestigious medical journals isn't necessarily true. And even some of the best physicians are typically unaware of this fact.

7. Study Money

As a general rule, scientific studies that draw the most financial support today are ones likely to generate profits for a company. Few companies are willing to sponsor research that could reveal bad things about their medicine, treatments, products, or chemicals. For every study that does unearth something negative, government and industry money is spent launching five more studies to negate it. Crucially important studies that don't stand to generate big money for someone are often bypassed. Or, if they do get funded, the results may not be well publicized in the press because nobody powerful is promoting them.

Around 2019, the fifth generation of cell phone technology, 5G, is being rolled out, and advocates around the globe are pointing to the possibility that it will compound the known but little publicized risks from cell phones such as infertility, depression, and other chronic disorders. For a report on my Sunday TV news program *Full Measure,* I interview one of the world's preeminent scientists on electromagnetic fields (EMFs), Professor Martin Pall, to find out if the concerns are valid.

Pall is quite the character. When we meet for my interview at my Arlington, Virginia, studio, he looks to be in his seventies and fit, wearing a navy blue suit and tie, and glasses, and sporting a healthy head of slightly disheveled gray hair. He's so concerned about the potential for health risks from EMF, he even wears special clothing to shield him from the rays. When I ask if he's wearing the fabric now under his suit, he unbuttons his dress shirt to show me his metal-fiber-embedded undershirt. He doesn't carry a cell phone and only uses a wired connection from his computer to a printer. "And so everything's wired, and I even have a shield over the screen so that I get less EMF from the screen," he explains.

"We know that the EMF impact the cells of our bodies, all the cells of our bodies, by activating some channels," Pall tells me. "And when they do that, they produce all kinds of effects. And those include neurological neuropsychiatric effects. They include reproductive effects, they include oxidative stress, which is involved in essentially all chronic diseases. So I'm deeply concerned about the situation."

Pall says EMF can cause chronic tiredness, anxiety, headaches, poor concentration, and worse. At least three expert medical groups have connected certain kinds of EMFs to cancer—particularly childhood leukemia. Pall is convinced 5G will be a long-term disaster for public health in ways nobody is closely monitoring. I begin to think, *Maybe he's not so much a character as he is wise to the scientific evidence.* I begin to wish I had one of those metal fiber shirts too.

As I dig into the research, I find Congress, federal agencies, and even the telecommunications industry that produces 5G all admitting there are no studies proving 5G is safe for humans. But the way the industry spins it, that means there's no proof of harm either. It's too new. *Let's just deploy it and see what happens,* they argue. Even the FDA, frequently in the pocket of industry, did concede some months earlier that there *should be* new research done on cell phone safety. The FDA urged the cell phone industry to do its own research to see if its products cause harm. What are the odds they can be trusted to expose the risks of their own multibillion-dollar technology?

At a congressional hearing in February of 2019, months after the FDA suggested to the industry that it research itself, Senator Richard Blumenthal asks wireless representatives if they've actually launched any studies. Each wireless official replies: No.

The potential harmful consequences of this reality cannot be overstated. As we've learned, in today's convoluted scientific landscape, studies that could expose important safety concerns either aren't done, or are funded by companies with the power to censor any results showing harm. Who's left to conduct crucial research that reveals the combination of factors causing epidemics of chronic diseases in America? What effect is pesticide-tainted drinking water having on the population when combined with chemicals in food, cell phone exposure, hormones in beef and milk, and the particular medicine an individual takes? These complex, multipart questions *are* answerable—if only someone were to try.

Even research that's funded with taxpayer money suffers from the same selective syndrome. The National Institutes of Health (NIH) has proven to be hopelessly political—and money-driven. NIH is tied closely to industry interests and uses its power to control which studies, researchers, and institutions do and don't get government funding for studies. What's worse, NIH scientists working on our dime are allowed to partner with private pharmaceutical companies and collect royalties for their inventions. They personally benefit through their relationships with outside corporations. It's a legally permitted conflict of interest that has no equal in private industry that I'm aware of.

Independent-minded scientists have told me there's important research they'd like to do, but they say they'd have to contort the framing of any grant proposal into a question the federal government and pharmaceutical industry want answered, or else they won't get a dime. Typically, these scientists say they don't bother to propose studies on matters that are off the narrative. There's no point.

Lessons Learned: Public health policies are driven and controlled through the granting or withholding of research funds. Some research that is most important to public health is not getting funded.

8. Crossing a Line

As I came to know what I know today, it baffled me that many in the media and the public automatically treat research and the word of health officials as if they're infallible. Never dishonest. Beyond reproach. For example, when watchdogs or consumer advocates raise reasonable questions about a medicine's safety, they inevitably get hit with public attacks. *What—so now you have a medical degree?* Or *Thanks, but I'll trust the word of the one who trained in medical school for seven years.*

That rhetoric is stoked by paid interests working to controversialize anyone who threatens a drug company's bottom line. Oddly, while many consumers understand that profit-driven corporations can commit the worst kind of unethical and in some cases criminal behavior—think Enron, Firestone, Ford, and Solyndra—they seem incapable of fathoming that pharmaceutical companies could be capable of the same. But a lengthy record proves that doctors, public health officials, and drug firms have gotten caught committing major bad acts that endanger patient health.

One need look no further than the long list of payments big pharmaceutical companies have made to settle criminal or civil complaints. Each settlement has stunning global health implications in terms of how many people were impacted. Each settlement should have prompted international stories and investigations that filled news space for months or years. But I'll bet you don't know much, if anything, about them. After you read the list below, and learn more about the widespread deceit or wrongdoing committed by so many players, you may become more skeptical, as you should, about the entire system that approves, prescribes, and defends what's on America's medicine shelves.

In 2023, Teva Pharmaceuticals admitted to taking part in three conspiracies and agreed to pay more than **$200 million** in fines to settle price-fixing charges related to a widely used generic drug for high cholesterol. It was said to be the largest penalty paid "for a domestic antitrust cartel" and ended a price-fixing and bid-rigging investigation of seven companies that netted a total of **$681 million** in fines.

In 2020 and 2021, numerous generic drug manufacturers paid **$400 million** in criminal fines in a lucrative, widespread price-fixing and kickback scheme.

In 2020, Purdue Pharma was criminally charged with marketing and selling opioids—powerful and often-abused pain-relieving medicine—when company officials knew the medicine was being diverted to drug abusers. Additional charges included paying kickbacks to doctors, and reporting misleading information. The company settled by shelling out more than **$8.3 billion** in penalties. The drugs involved were Butrans, Hysingla, and OxyContin.

Also in 2020, Novartis—the same company from our frog professor story—agreed to pay more than **$591 million** over allegations of paying bribes and kickbacks to get doctors to prescribe ten drugs: Lotrel, Valturna, Starlix, Tekturna, Tekturna HCT, Tekamlo, Diovan, Diovan HCT, Exforge, and Exforge HCT. According to prosecutors, "For more than a decade, Novartis spent hundreds of millions of dollars on so-called speaker programs, including speaking fees, exorbitant meals, and top-shelf alcohol that were nothing more than bribes to get doctors across the country to prescribe Novartis's drugs." The government alleged that Novartis sales reps selected high-volume prescribing doctors to serve as paid "speakers" at sham events and induce them to write more Novartis prescriptions. At the same time, the company agreed to pay more than **$51 million** in a settlement over allegedly using foundations in an illegal scheme to sell more of its multiple sclerosis drug Gilenya and its cancer drug Afinitor. That settlement also involved alleged kickbacks to doctors. Separately, Novartis agreed to pay **$345 million** to settle charges that it bribed medical professionals, hospitals, and clinics in Greece to prescribe Novartis products and use its subsidiaries' surgical products and then falsified records to cover up the bribery.

GlaxoSmithKline settled criminal and civil charges in 2012 with a

$3 billion payoff over allegations of misbranding, deceptive marketing, paying kickbacks to doctors, failing to report safety data, and illegally promoting drugs for unapproved uses. The drugs involved were Advair, Flovent, Imitrex, Lamictal, Paxil, Valtrex, Wellbutrin, and Zofran.

A few other big ones in the past fifteen years, starting with the largest in terms of fines, include: **$2.4 billion** paid by Takeda Pharmaceutical in 2015 to settle charges that it hid bladder cancer risks and deceptively marketed Actos; **$2.3 billion** paid by Pfizer in 2009 to settle charges that include deceptive marketing, unlawful promotion, and paying kickbacks to doctors concerning Bextra, Geodon, Lyrica, and Zyvox; more than **$2.2 billion** paid by Johnson & Johnson in 2013 to settle cases claiming deceptive marketing, kickbacks, and illegal promotion of Invega, Natrecor, and Risperdal; **$1.5 billion** paid by Abbott Laboratories in 2012 for similar allegations about its drug Depakote; more than **$1.4 billion** paid by Eli Lilly in 2009 for Zyprexa; **$950 million** paid by Merck in 2011 for Vioxx; **$875 million** in 2001 from TAP Pharmaceutical for Lupron; **$762 million** in 2012 from Amgen for Aranesp, Enbrel, and Neulasta; **$775 million** from Bayer and Johnson & Johnson in 2019 for Xarelto; **$520 million** by AstraZeneca in 2010 for Seroquel; and **$360 million** in 2018 by Actelion for Opsumit, Tracleer, Ventavis, and Veletri.

An October 2023 analysis puts pharmaceutical settlements and penalties since 2010 at more than $80 billion. Johnson & Johnson tops the list with more than forty-five violations and $24.5 billion in penalties. Between 2003 and 2016, 85 percent of top pharmaceutical firms paid financial penalties for illegal activities. The highest penalties compared to the percentage of revenue were paid by Allergan, GlaxoSmithKline, Schering-Plough, and Wyeth. The firms with the greatest variety of alleged illegal activities were Bristol Myers Squibb, GlaxoSmithKline, and Merck. The analysis concludes, "Given the scope and nature of the illegal activities involving financial penalties, physicians and regulators should exhibit vigilance over the activities of large pharmaceutical firms." *How can our system be so broken that these violations are so commonplace?* So many drugs and companies are implicated, it would be logical to conclude that lying, deceptive marketing, hiding safety risks, and paying kickbacks are simply part of doing business in the pharmaceutical industry. It shouldn't escape notice that for every charge of bribery and kickbacks, it implies there were plenty of crooked doctors willing to be bought. And even as the medical establishment tries to elevate vaccines into a revered category of their own, never to be questioned, when you realize the biggest offenders on the legal violations list include vaccine makers such as GlaxoSmithKline, Pfizer, and Johnson & Johnson, is it rational to believe that they would never mislead when it comes to their multibillion-dollar vaccines?

The Department of Justice (DOJ) churns out self-congratulatory press releases about hefty financial penalties it wins from pharmaceutical companies. At the same time, government prosecutors seem unfazed by the fact that the bad behavior continues year after year. And there aren't many DOJ announcements indicating corporate executives are getting prosecuted. That helps explain why the wrongdoing continues. Companies make so much money from their illicit activities that it's worth the risk of getting caught. Even if they eventually pay a fine, it's still a net gain. According to a May 2024 analysis by the watchdog group Public Citizen, from 1991 to 2021, the government won payments of $62.3 billion in penalties from pharmaceutical companies. But the group observed, "This total amounts to a small percentage of the $1.9 trillion in net income made by the 35 largest drug companies during just 19 of those 31 years (2000–2018)." Can you imagine what a game changer it would be if even one high-profile Big Pharma CEO were prosecuted for alleged fraud? Or if companies engaging in deceptive practices were barred from having new drugs considered for approval for ten years? As it stands, government prosecutors get to brag about getting hundreds of millions of dollars in fines with no tangible benefit to individual consumers, as the offending pharmaceutical executives often go happily about their business.

While major companies are capable of great wrongdoing, there is ample evidence that individual researchers too can allegedly skirt rules and laws. There are too many instances to quantify here, but watchdogs report that from 1992 to 2012, taxpayers shelled out $58 million in grant money to scientists whose papers were later retracted due to misconduct. *And those are only the ones who got caught.*

I reported on one seminal case while I was at CBS News. It involved a research doctor who potentially hurt millions of people: Dr. Eric Poehlman at the University of Vermont School of Medicine.

Dr. Poehlman got $2.9 million in federal grants and published hundreds of medical articles on obesity and menopause—before he got caught fabricating data. Who exposed him? A young research assistant who'd previously looked up to Dr. Poehlman as a hero. One day, after an assignment crunching data for Dr. Poehlman, the assistant says he observed Dr. Poehlman presenting results that were flipped: the opposite of what the data actually showed! When the assistant raised the issue with the good doctor, he discovered the misinformation was no accident. Dr. Poehlman had no intention of correcting his errors.

The research assistant tried to report the fraud to some authorities inside the college, but to his chagrin, he says nobody wanted to hear it. His own future as a physician was threatened if he persisted in raising questions. He was told, *We'll make sure no medical school accepts you.* Still, he escalated his concerns until someone at the university finally listened and responded. *How many young students would have done the same? How much fraud goes undetected?* In 2006, Dr. Poehlman became the first academic researcher sentenced to prison for fraudulent studies. The scandal resulted in corrections and retractions of ten scientific studies. It's unlikely that all the doctors who read Dr. Poehlman's menopause and hormone replacement therapy research in medical journals were aware that he falsified data. Long after the misconduct was flagged, I saw that Dr. Poehlman's work was still being cited by other researchers.

Many years later, I wondered if the research assistant who exposed Dr. Poehlman's deceitfulness ever got to medical school. I found him successfully practicing as a doctor. I was glad he'd made it—an honest, informed voice amid a sea of incompetence and corruption. I asked him to do a new interview about research ethics for my television show. He said he'd rather not rock the boat. I think he should have been proud to have stood up to corruption. But it seems to me that he knew that the system in which he now operates might view him as a snitch and not treat him kindly.

In 2001, I investigated the heartbreaking case of an infant named Gage Stevens. Gage died while participating in a study for a heartburn drug called Propulsid, made by Johnson & Johnson. The more I uncovered about the lead researcher's conduct, the more outlandish the story became. Unbeknownst to Gage's parents, Propulsid had been linked to dozens of deaths, including that of at least one other baby, before Gage even got his first dose! My investigation revealed that the study consent form presented to Gage's parents falsely stated Propulsid was "approved by the FDA" for children. In fact, the Food and Drug Administration had repeatedly rejected the drug for pediatric use. There were eerie similarities to the Baby Oxygen story that I would report on years later.

After my Propulsid report about Gage's death aired on the *CBS Evening News*, the FDA publicly faulted the study's leader, Dr. Susan Orenstein, for violating federal regulations and good clinical practices. Among other problems, the FDA identified poor recordkeeping by Dr. Orenstein, and a consent form that was coercive and didn't clearly state the risks. Researchers have been accused of misleading study subjects on consent forms for decades. The FDA also said Dr. Orenstein failed to report all serious adverse events her study patients suffered, including another baby who developed "symptoms of severe intolerance—screaming spells, crying, pulling legs—after each dose." Even worse, I obtained evidence suggesting Gage never even had the disorder that supposedly made him eligible to be part of the Propulsid study. Dr. Orenstein originally did a biopsy and claimed to Gage's parents that his esophagus was inflamed from acid reflux, so he qualified to be part of the drug experiment. But the actual test results told a different story. The biopsy was negative. There was "no evidence of . . . significant inflammation." His parents had been tricked. Gage should never have been in the ill-fated experiment. Twenty-two years later, I checked and could find no record that any meaningful discipline had been given to Dr. Orenstein. Under the FDA's newly loosened informed consent rules I spoke of earlier, researchers like these now have free rein to declare that their study poses "minimal risks" and, therefore, not disclose all of the risks to study subjects. You know, for the greater good.

More recent misconduct allegations were lodged in September 2022. Three medical journals launched investigations into a group of heart studies led by Temple University researchers involving the blood thinner Xarelto. They claimed Xarelto could have healing effects on hearts. The research was funded by you, the taxpayers, through the National Institutes of Health and Janssen Pharmaceuticals. However, the findings were retracted amid various investigations, and the researchers were questioned for alleged data manipulation.

Lessons Learned: Blind trust cannot be given to people simply because they are researchers, or have a medical degree or license. Fraud, misconduct, and wrongdoing in the pharmaceutical and medical industries are well established.

9. Intertwined Interests

There's a vast network of unseen, intertwined interests all designed to spin and influence us on science and health matters. Seemingly independent entities are secretly linked. They work together to declare a potentially dangerous drug "safe." Or debunk a study that raises concerns. Or smear researchers and reporters who ask legitimate questions. Or delay adding warnings to a medicine. Or postpone its ultimate withdrawal to give time for more profits to be collected.

At first, the connections were invisible to me. But my cognitive dissonance led me on a quest to fill in the blanks. Much as I had wondered why reputable physicians would blindly defend drugs that had safety issues, I began to ask why political figures, federal agencies, nonprofits, consumer groups, and other organizations seemed blissfully blind to health concerns about medicine. Were they masking behind-the-scenes relationships with drug companies, as the physicians had?

An "Aha!" moment came when I was first investigating links between vaccines and autism. To my surprise, I learned that the connection between vaccines and brain injuries, including autism, turned out to be well documented in scientific literature and in court. I began breaking quite a bit of related news on the topic. But it's not because I was such an amazing detective. It's because I approached my research with the commonsense practicality of an outsider rather than a medical reporter. I was relatively immune to the heavy spin of government and industry insiders who routinely managed to convince beat reporters to look the other way or dismiss studies that provide reason for concern.

Minutes before one of my vaccine-autism reports was to air on the *CBS Evening News,* I'm preparing on the small remote set of the Washington, DC, bureau to give a live introduction to the story. The hotline from the New York main office that rings directly into the Washington bureau lights up. It's for me.

"Why is some group called 'Every Child By Two' supposedly fronted by Rosalynn Carter calling me about your story?," executive producer Jim Murphy asks. It turns out a spokesman for the nonprofit group "Every Child By Two" had just called the network in New York to try to prevent my story from airing. In order to get the call through to top newsroom officials, the spokesman had evoked the name of former First Lady Rosalynn Carter, a figurehead for the group.

"I have no idea," I reply. I wondered how Every Child By Two knew we planned to air a story that night. *Why is anybody calling about a story that hasn't even aired yet?*

"Your story's solid, right?" Murphy asks, already knowing it's been through multiple reviews, including a legal review.

"Yes," I assure him. There's not a sliver of doubt.

At the time, Murphy had been assigning me to do tough investigations into pharmaceutical controversies and scandals, which impacted millions and were of extremely high interest to viewers. There were enough to do one a week for the rest of my career. But unknown to me, Murphy was beginning to catch flack for our stories from the CBS advertising department and beyond. In retrospect, I realize how brave Murphy was in moving forward with the assignments, considering the powers that we were up against. Today that kind of bravery is rarely found at major news organizations.

After I finish my live shot, I decide to look into Every Child By Two. It purports to watch out for the best interests of babies, making sure they get all of their recommended vaccinations by age two. It doesn't make sense to me that a group like that would try to block a factual story about vaccine concerns. *Shouldn't they be more interested than anyone in vaccine safety?*

I get my answer when I locate the nonprofit's IRS tax form, its Form 990. And there it is. The bulk of the funding for Every Child By Two came from vaccine maker Wyeth, with a Wyeth spokesman listed as treasurer. The way I see it, Every Child By Two is no more than a front for the vaccine industry under the guise of a nonprofit led by Rosalynn Carter! It leaves me to conclude the organization isn't interested at all in exposing vaccine safety issues; its mission is to quash them.

Meantime, hired guns for pharmaceutical interests flood me and CBS News with emails, phone calls, and requests for meetings. The spokesman for Secretary of Health and Human Services Tommy Thompson calls the CBS News Washington bureau chief to exert pressure in an attempt to discredit and kill the whole line of vaccine safety stories we're doing. Pharmaceutical company lawyers set up secretive meetings with CBS officials in New York. Pharmaceutical interests and their high-priced PR firms contact CBS executives to complain. Every Child By Two is on that same team.

The practice of cloaking financial interests isn't limited to Every Child By Two or even the pharmaceutical industry. An investigative producer once joked to me that when a nonprofit calls itself one thing, you should assume its goal is the opposite. There's some truth to be found in the joke. When there are emerging safety questions about a drug, the drugmaker knows that reporters search the Internet for victims to interview. The drugmaker knows that injured patients network online. So the drugmaker, or its proxy, starts its own "patient advocacy" group providing real-life "patients" who will side with the drug. The "patient advocacy group" posts blogs and articles praising the drug, blaming other factors for any illnesses, dispelling the idea of drug risks, and making the case for how badly patients will suffer if the medicine is removed from the market. And they're happy to do media interviews. Some of the top patient or disease advocacy organizations you routinely hear from in the news fall into this category. Maybe it's somewhere in the fine print, but it's not usually made obvious that the group is a front for a pharmaceutical interest. *"Patients for Quiloxone Safety"? Brought to you by the makers of Quiloxone, of course.*

Toward this point, a 2023 Yale University School of Medicine analysis finds that 74 percent of the "highest-revenue patient advocacy organizations" have board members or senior leadership tied to Big Pharma. Stating the obvious, the analysis concludes that the interwoven relationships raise questions "about industry's influence on these organizations' patient education, policy recommendations, and treatment guidelines." Coauthor Shamik Bhat observes, "What was particularly concerning . . . is that some of these organizations had . . . leadership . . . who . . . currently were actively working for industry, including some CEOs or executive directors."

One example is the American Cancer Society (ACS). The Yale report found that the American Cancer Society's chief executive officer, Dr. Karen Knudsen, is also a member of Genentech's Scientific Resource Board. Genentech markets and develops cancer drugs and can benefit if the ACS endorses or promotes its products. And the ACS also stands to gain from the endorsement. For example, Genentech has given millions to the American Cancer Society for partnerships that could put its drugs into the hands of more patients. Also, according to the Yale analysis, top executives at three other patient advocacy groups—Michael J. Fox Foundation for Parkinson's Research, the Cancer Research Institute, and the Foundation Fighting Blindness—were serving simultaneously on pharmaceutical boards: Pfizer, Coherus BioSciences, and Opus Genetics.

Another conflict of interest example is the Academy of Nutrition and Dietetics, and its website eatright.org. The Academy is the largest group of nutritionists and dieticians in the world, advising untold numbers of consumers and professionals on what to eat and what not to eat. A five-year investigation by the watchdog group US Right to Know found the Academy of Nutrition and Dietetics was taking millions from the very industries blamed for much of our poor diets and chronic diseases: pharmaceutical companies, agribusiness companies, and companies that make sugary sweets and highly processed foods. They include Conagra, Abbott Nutrition and Laboratories, Pepsi, Coke, Hershey, General Mills, Kellogg, the National Confectioners Association, and Bayer CropScience. For its part, the Academy of Nutrition and Dietetics says it has rules that maintain strict firewalls between corporate donations and policies, and that their sponsors have no influence on the advice they give.

The industries that seek to control the national discussion on our health know how to pull strings in many other ways. They may work through PR firms or other third parties to indirectly fund propaganda. Or maybe they get legislative favors from Congress by starting a charity where the charitable cause is a member of Congress.

Take the case of Stephen Buyer, a Republican from Indiana, whom I investigated at CBS News. As a member of Congress, Buyer used a $25,000 donation to start a nonprofit called the Frontier Foundation to fund education scholarships. It was later revealed that the Frontier Foundation collected more than $800,000 over six years but didn't give out a penny in scholarships. The obvious questions: Who provided that $25,000 donation, and what was the true purpose of the Frontier Foundation?

Representative Buyer declined my interview requests and literally jogged in the other direction when he saw me and my cameraman looking for him in the halls of Congress to ask questions. After a few days of this drill, Buyer's office told me that if I would stop following him around, he'd give me a brief on-camera interview. By the time we sat down for the interview in his office, I had traced every donation on record to his Frontier Foundation. It turned out to be big money from the pharmaceutical, tobacco, and telecommunications industries. By examining the Congressional Record and cross-referencing dates, I found that each donation to Buyer's "charity" matched up to a specific congressional action Buyer took.

Within days of advocating for initiatives that benefited the contributing industries, he got donations from them.

In our on-camera interview, I ask Buyer to tell me who put up the original $25,000 to start the nonprofit. He says he "can't recall." I don't find that very credible. *You don't remember who gave $25,000 to start a nonprofit in your name?* I then read to Buyer from an incriminating list I've compiled. For example, after the Frontier Foundation received hundreds of thousands of dollars from pharmaceutical interests, Buyer took up a number of pharmaceutical industry priorities and sponsored bills the industry supported. After RJ Reynolds and other tobacco interests gave generously, Buyer opposed a bill giving the FDA authority to regulate tobacco. Instead, he sponsored an RJ Reynolds–supported alternative.

So what did the Frontier Foundation spend money on if not education scholarships? Much of it reportedly paid for golf tournaments in the Bahamas and other fundraising events frequented by industry lobbyists and Buyer himself. Foundation funds also paid for Buyer's travel to and from these events, where industry lobbyists had direct access to his fullest attention.

As my line of questioning continues, Buyer tells me it's unfair to imply that an independent man like him would act on behalf of the interests that are donating large sums to his charity. I have more questions, but our brief interview time draws to a close. He announces he has someplace to go, stands up and pulls off his microphone, weirdly grabs a nearby bottle of hand sanitizer, and tosses it into my lap as he strides out the door. We lead the *CBS Evening News* with the story. Shortly after my report, Buyer acknowledges the startup money for his nonprofit came from the pharmaceutical industry and announces he's retiring from Congress. He goes on to work for a drug company, a tobacco company, and a telecommunications firm. In 2023, he gets convicted of insider trading of telecom stock.

Lessons Learned: The pharmaceutical industry blankets the information landscape in ways that are invisible to us and made to appear grassroots in nature. The word of news organizations, nonprofits, or government agencies and officials cannot be accepted without question. A pharmaceutical agenda is often driving the public narrative.

10. Antiperspirants and Cancer

In 2005, I begin looking into why warnings about kidney damage were recently added to antiperspirant labels. But I stumble across an even bigger story. An FDA source I'm talking to about the kidney warnings discloses to me that multiple studies have linked antiperspirants to *breast cancer.* I'd never heard of such a thing. It seems immensely important. The FDA, the official tells me, has been wrestling with the cosmetics and antiperspirant industry for years over adding a breast cancer warning to antiperspirant labels. So far, the industry has successfully beaten back the idea.

I eventually produce a groundbreaking two-part report for CBS News that reveals numerous studies have linked aluminum in antiperspirants to breast cancer, and that the FDA website contains false information on the topic. On the web at the time, the FDA mimics the industry stand and calls the antiperspirant link to breast cancer "false . . . scary stories." But when I press agency officials on studies that say otherwise, they shift their story. They concede that a breast cancer link has not been ruled out, and say more studies should be done. In a statement, the agency writes,

> FDA is aware of concerns that antiperspirant use (in conjunction with underarm shaving) may be associated with increased risk of developing breast cancer. FDA continues to search scientific literature for studies examining this possible adverse drug effect. Unfortunately, there are many publications that discuss the issue, but very few studies in which data has been collected and analyzed. Overall, the studies [containing data] are inconclusive in determining whether antiperspirants, in any way, contribute to the development of breast cancer. FDA hopes that definitive studies exploring breast cancer incidence and antiperspirant use will be conducted in the near future.

Did the National Institutes of Health carve out a slice of its research billions to provide "definitive studies" about a product used by hundreds of millions? *Nah.*

The antiperspirant story also gave me an education in a conflict of interest I knew nothing about. When I first asked the cosmetics and antiperspirant industry for an interview, representatives referred me to the American Cancer Society for assurances that antiperspirants can't cause cancer. I found that a little strange. Indeed, when I contact the American Cancer Society, the director of content, Dr. Ted Gansler, is quick to rule out a breast cancer link. But when I ask him about studies to the contrary, I'm shocked to learn that he's unfamiliar with them. He asks me to fax him the studies. Once he has the scientific evidence in hand, instead of addressing it, he deflects. He answers each question I ask about the breast cancer risk by repeating that women would be better off getting mammograms than worrying about other "small risks" like antiperspirants. *Why would the American Cancer Society be giving definitive opinions debunking the link when their experts were unfamiliar with the relevant science? Why do they keep talking about mammograms instead of addressing the question at hand?*

By now, I'd learned to follow the money, especially when a response doesn't seem to add up. I take a guess and ask Dr. Gansler if the American Cancer Society receives money from the cosmetics industry that represents antiperspirant makers. He bristles. I press. He finally acknowledges that, yes, it does accept industry money, but he won't tell me how much or even give a range. He further says that he won't give an on-camera interview for my TV news story unless I agree not to ask him about the funding. In my resulting investigation, I include the American Cancer Society's position that women should get annual mammograms and not worry about breast cancer and antiperspirants. I also report that the American Cancer Society acknowledges receiving a "small amount of funding" from the cosmetics industry.

That disclosure, in my fact-based report, sparks an angry letter from Dr. Gansler to my executive producer at CBS. To this day, the letter remains an exemplary lesson in how nonprofits often tow the line for industries that support them and push back against factually accurate news reports they don't like. In his complaint, Dr. Gansler calls my report "sensational and misleading" and says that instead of dispelling a "common misconception," I furthered a "baseless and scary rumor." He says that my disclosure about the American Cancer Society getting industry funding was "clearly used to imply that donations from the industry colored our interpretation of the evidence." He added, "This is inaccurate and offensive."

CBS publishes both Dr. Gansler's letter and my response (on our website). In my response, I point out that Dr. Gansler had referred to his group, the American Cancer Society, as "the top source of credible, science-based information," yet he'd given a definitive position on something while unfamiliar with the latest research. I recount,

> When we first spoke on the phone, your organization mistakenly stated that the antiperspirant breast cancer link had been widely disproven and asked if there was "something new." I told your organization there were several recent studies, including the one by Dr. Kris McGrath. Your organization told me you were unfamiliar with such studies. Therefore, as you know, I provided Dr. McGrath's study to you for your review prior to our interview. . . . The idea of a cancer link might indeed be "scary" as your letter states, but it is certainly not "baseless." Just because people might find an idea scary doesn't mean the truth should be withheld. . . . One has to wonder why the American Cancer Society . . . would go to such pains to dispel a potential link between antiperspirants and breast cancer when no such link has been definitively dispelled; and why your organization seems so uncomfortable with the public being told that it's an open question rather than a closed issue. Instead of being in tandem with the antiperspirant industry, one might have expected the American Cancer Society to support public dissemination of all information regarding potential cancer risks—especially in cases where the public can choose to take steps to reduce exposure that could add to risks . . . The public has the right to know the full truth and make its own decisions.

Two additional insider footnotes: All of this occurred during a time when CBS began capitulating to lobbying and pressure from industry interests. Some of it seemingly from within the network. Prior to my two-part breast cancer investigation airing, the executive producer told me that our own CBS medical reporter and her producer in New York had approached him to try to get him to cancel my story.

"What reasons did they give?" I ask the executive producer.

"I don't know," he says. "[The medical producer] gave me a piece of paper with their objections written on it. I tore it up into tiny pieces and put it back in his hand."

More attacks come the night after part one of my two-part antiperspirant–breast cancer investigation airs on the *Evening News*. A top CBS News executive who often deals with corporate matters above the news division calls me at home with an unprecedented request. She argues that "part one was so good, there's really no need to air part two." She wants me to withdraw part two, scheduled to air the following night. (It won't be the last time that a CBS manager will claim that my reporting is just so good, so thorough, that there's no need to do more of it.) I calmly tell the executive that part two, explaining the difference between antiperspirants and safer, aluminum-free deodorants, is an entirely different report, and of high interest. When we hang up, I call my executive producer and ask if we're moving forward with airing part two. With his backing, we did.

There would be future battles that we would not win. Some of them involve my reporting on what one government advisor privately called a "third rail," not to be touched. Vaccines.

CHAPTER 6

The "Third Rail"

Vaccines

Has your doctor warned you about potential side effects each time you or a loved one got a vaccine? Under established ethics guidelines, you should be informed about everything from the risk of paralysis to brain damage and death, depending on the vaccine. If you weren't told of these risks, then did you truly give your *informed* consent to be vaccinated?

It was a long time ago, but I recall being surprised when I first discovered public health officials using official channels to perpetrate vaccine misinformation. I was startled by their manipulation of news and information. A large part of their propaganda campaign revolves around mislabeling accurate facts as "disinformation," smearing certain scientists, and falsely attacking people as "anti-vaccine."

Let me pause to state something that should be obvious. It's not "anti-vaccine" to ask questions, research, or report about vaccine safety. In all my years of investigating topics, this one stands alone in terms of the magnitude of orchestrated pushback it draws. When I broke international news about deadly rollovers of Ford Explorers outfitted with Firestone tires, nobody suggested I was "anti-car" or "anti-tire." That would be absurd. When I uncovered fraud at the Red Cross involving 9/11 donations, nobody suggested I was "anti–Red Cross," "anti-charity," or "anti–9/11 donations." When I investigated other drug safety issues, nobody considered me to be "anti-medicine." If I were to talk about studies showing that some people are allergic to penicillin, it wouldn't make me "anti-penicillin." Ask yourself why the game changes when it comes to vaccines. The fact that people feel compelled to say, "I'm not anti-vaccine . . ." before making perfectly grounded statements or asking rational questions speaks to the success of one of the most influential propaganda movements of our time.

One of the cruelest things our government does is smear the poor parents of vaccine-injured children. For daring to speak publicly about what happened to their loved ones, these parents are attacked by public health officials and held up to ridicule by the media. Parents have told me the government has proven vindictive and they fear if they don't keep their mouths shut, the government will take back payments awarded by courts to care for their vaccine-injured children. They're literally bullied into silence. I think part of the reason why vaccine interests are so intent on neutralizing parents is that parents are the most important and credible spokesmen when it comes to vaccine safety. By definition, they weren't "anti-vaxxers." *They vaccinated their children.*

All medicine has side effects. But in today's manipulated information landscape, efforts to learn the most about side effects of vaccines, products given multiple times to virtually every American, are actively discouraged. We're made to think that questions are not even to be raised. This is the antithesis of good science and public health. Without risks being addressed, our national vaccine program is neither as safe nor as effective as it could be. The fact that even medical professionals who should know better treat vaccine safety as a third rail not to be touched makes no logical sense and serves as an important giveaway that a commanding narrative is in play.

15 Eye-Openers

Many books could be written as to why federal agencies and public health officials entrusted with our well-being are part of the global vaccine propaganda campaign. I'll succinctly highlight fifteen eye-openers that I discovered through extensive research. These examples helped shape my reporting and how I approach current medical stories.

Prior to my years of research, I wouldn't have believed any of these examples were true! When I eventually published news stories on these topics, it was usually after months or years of investigation. I wanted to make certain I was relying on accurate research and credible sources—precisely because there's so much propaganda claiming the opposite of what turns out to be true. Also, please note that due to a broad misinformation campaign, if you google these topics, you'll be bombarded by "fact-checks," news articles, and medical citations that "debunk" the truth. We'll dissect that a bit later. For the sake of brevity and clarity, I'm using examples here that cannot be legitimately disputed. And I'm including some simple references and citations, in case you want to learn more.

1. **DPT**: The shot so many of us got as children to protect against diphtheria, pertussis, and tetanus, the DPT triple vaccine, caused seizures and brain injuries in some children, according to medical testimony and scientists. In 1994, a US Institute of Medicine (IOM) committee agreed that "the balance of evidence is consistent with a causal relation between DPT" and brain damage. Numerous lawsuits awarded up to $5 million to each injured child. This was a precursor to the wave of "autism" that some researchers say is simply the name we've given to specific types of brain damage. In response to the 1994 DPT brain injury determination, the government recommended a revised formulation in 1996 hoping it would be safer for children (though there was no evidence to say one way or another): DTaP. Like all childhood vaccines, the DTaP version was never tested in combination with other shots babies get. If you google vaccine safety, it returns an impressive body of propaganda falsely insisting that no products have been more rigorously tested for safety and effectiveness than vaccines. In fact, there are no public studies showing the cumulative, real-world impact of the CDC's recommended childhood vaccine schedule on any individual. In

2013, the powerful Institute of Medicine (IOM) weighed in on this research deficit. The IOM was a unique group of experts, chartered by Congress, intended to provide objective medical advice on health matters. In the view of critics, the IOM frequently acted largely as an arm of the government and pharmaceutical industry on some controversies. In 2015, it changed its name to the National Academy of Medicine. But in 2013, the IOM actually admitted the cold, hard reality. "Most vaccine-related research focuses on the outcomes of single immunizations or combinations of vaccines administered at a single visit," wrote an IOM committee. "Thus, key elements of the entire schedule—the number, frequency, timing, order, and age at administration of vaccines—have not been systematically examined in research studies." So, did the National Institutes of Health or our public health experts snap to and close this glaring research gap? *Nah.* Billions upon billions of tax dollars are spent on research, but not a penny to study the exposure nearly every American baby gets.

Read about the government's findings on DPT at this link: https://www.ncbi.nlm.nih.gov/books/NBK225455/

Read an example of DPT vaccine injury here: https://www.uscfc.uscourts.gov/sites/default/files/opinions/Almeida2.pdf

Read the National Vaccine Information Center's summary of vaccine risks at this link: https://www.nvic.org/vaccination-decisions/know-the-risks

2. **Vaccine Court**: By the early 1980s, DPT vaccine makers were already facing so many brain damage lawsuits, they got together and threatened to stop producing the shots unless Congress protected them from liability. As a result, in 1986, Congress passed the National Childhood Vaccine Injury Act. The Act did something unheard-of. It created a special vaccine court under the United States Court of Federal Claims to handle vaccine injuries. The court operates under unique rules. The burden of proof required of a vaccine-injured patient is unrealistically high, and the statute of limitations is unfairly short, according to victims' advocates. Patients are denied the normal "discovery" process

that could show what the drug companies knew and when they knew it. By the time many parents suspect their child has suffered autism or another vaccine injury, it's often too late for them to obtain the contemporaneous documentation the court requires. A majority of claims are denied. Though most parents aren't aware vaccine court exists, the relative few who do find it, and have the required documentation, have received a cumulative $5 billion for more than 10,600 claims.

Incredibly, under this sweetheart arrangement for the industry, our own Department of Justice defends vaccine makers. This oddly pits taxpayer-funded government against injured patients and exempts the guilty party: vaccine makers. And when damages to victims *are* paid, the money doesn't come from the culpable vaccine makers. It comes from a fund built with fees charged to patients on each vaccine dose they receive. In other words, vaccine makers don't bear the cost of their products' damage—we do. Vaccine court has unwittingly created a database of thousands of self-identified, injured patients whose families would be happy to be studied so that risk factors for vaccine injuries could be identified and mitigated. Yet the government leaves that gold mine of data untouched.

Read more at the government court website here: https://www.uscfc.uscourts.gov/vaccine-programoffice-special-masters

Vaccine injury compensation data can be found at this link: https://www.hrsa.gov/vaccine-compensation/data

3. **One-Two, Switcheroo:** America's stunning spike in autism cases directly coincides with the dramatic uptick in childhood vaccines. By 2019, most children were getting three times as many vaccinations as kids got in 1983. By the early 2000s, large groups of autism claims were finding their way to vaccine court. And the vaccine industry, with help from the government, began a stepped-up PR strategy to avoid what threatened to become a monumental financial disaster.

   By way of background, studies and court cases today document many links between vaccines and instances of autism, both suspected and firmly proven. The variability in mechanisms and side effects makes it easy to throw the public off the scent. Vaccines aren't the only "cause" of autism. And "autism," say many experts, is simply a name assigned to a group of confusing and disparate symptoms surrounding encephalitis or brain damage. How vaccines cause or trigger autism isn't one simple, identifiable thing. Some scientists blame the preservative known as thimerosal, which contains toxic mercury. Others believe it's the live virus component of the MMR vaccine, or the combination of mercury-containing shots given near the same time as shots containing actual live viruses. Some court cases have found damage is caused by the immune assault on vulnerable kids and adults when they get multiple vaccines. Some studies have fingered additives or "adjuvants" such as aluminum that are supposed to make vaccines work better. Maybe it's the administration of vaccines to a child who's immune-suppressed or whose immune system is challenged by illness. You'll never read about this on CDC's website, but the government has conceded that shots triggered autism in a child who had an undiagnosed vulnerability: a disorder involving her mitocondria, the structures that provide energy to our cells. Vaccines given to kids born with a condition called tuberous sclerosis have also resulted in autism, and families have gone to court and won payments for that. Some scientists say the whooping cough or pertussis component of the triple DTaP shot can be problematic. The vaccine connection to autism could be any single one of these factors or a combination, depending on the child. And if that's not confusing enough, many other illnesses besides autism have spiked in an alarming way and are linked to vaccines, including immune-related disorders such as juvenile diabetes, Crohn's disease, postural orthostatic tachycardia syndrome (POTS), arthritis, multiple sclerosis, Graves' disease, Addison's disease, lupus, psoriasis, inflammatory bowel disease, celiac disease, skin rashes, and more. As you can see, it's about as easy to grasp on to as a warm bowl of Jell-O.

In any event, by 2002, so many parents were filing autism

claims, the vaccine court devised a unique plan to address them. Rather than litigate thousands of cases one by one, the court decided to hear out two specific theories. Even amid the autism epidemic, most ordinary people knew little about this giant set of cases.

The first theory to be tested was that the triple dose measles-mumps-rubella (MMR) vaccine containing live measles virus somehow interacts with mercury-containing vaccines, such as the triple dose diphtheria-tetanus-pertussis shot (DTP), to cause autism. Following the MMR DTP vaccine combo, many parents watched their children regress from healthy, normal development into a severely autistic state. The second theory the vaccine court planned to test was that mercury-containing vaccines alone can cause autism.

The process to litigate autism was given a name: the Omnibus Autism Proceeding. Six representative cases were chosen to test the two theories. They were heard by a type of appointed judge called special masters. If vaccines were found to cause autism in any or all of the test cases, then other autism cases with a similar set of facts would be automatically presumed to be vaccine injuries and compensated accordingly. The result promised to either quell growing questions about vaccine safety and government health policies or open the world's eyes to a hidden world of incompetence, conflicts of interest, and corruption.

Either way, it would be earthshaking.

I'd been learning about the vaccine-autism controversy for several years as a CBS News investigative correspondent. This scandal was proving eerily similar to others I'd investigated involving government, corporations, money, whistleblowers, and power. It involved subterfuge, dishonesty, and distractions. And always more than meets the eye.

I discovered that, despite insistences to the contrary in mainstream medicine and media, there's an impressive body of convincing evidence pointing to vaccines as triggering autism in some children. Numerous scientists assured me it's entirely possible for research to identify the vulnerabilities that make an individual child susceptible to vaccine injury, and to correct for them. These same scientists also told me that establishment medicine—from the CDC to some inside the Institute of Medicine (IOM)—had made a group decision to pretend there was no issue. In fact, they schemed to demonize those trying to expose the truth. The altruistic explanation for why they would be willing to mislead the public on such a grand scale? For the greater good. If public trust in vaccines were chiseled away by the truth, then preventable infectious diseases—they claimed—would make a major comeback, and we'd have a public health disaster on our hands. The nonaltruistic explanation, which I came to believe is probably closer to the truth, is that industry interests so permeate our health institutions, they're able to control narratives to protect their multibillion-dollar cash cows.

The Biggest Cover-up

Years after the Omnibus Autism Proceeding began, it's finally nearing a decision. I'm in my office at CBS News in Washington, DC, when I'm contacted by an impeccable, firsthand source connected to the vaccine court. We'll refer to him as James Howard (not his real name), to protect his identity. Howard has provided me with valuable background over a period of time, helping me become conversant on a variety of complex vaccine safety issues, how vaccine court works, and how much is at stake with the Omnibus cases. He is fiercely pro-vaccine. Yet, after becoming intricately familiar with many vaccine injury cases, he's also a thoughtful realist in terms of the damage they can do. On this day, he invites me to his office for an in-person meeting.

When I arrive, Howard delivers the bombshell. In the special masters' upcoming court decisions, he tells me, they'll announce at least one of the children in a test case *did* get autism as a result of vaccination. *Am I hearing this correctly?* I think.

"What does that mean?" I ask, for clarity.

It means, Howard tells me, that thousands more autism claims will be unleashed. Catastrophe will be visited upon the trust fund that pays families of the vaccine-injured. It will quickly go bankrupt. An entirely new way to replenish it will have to be devised. Unnamed government officials have already been briefed, he implies. The decision will rock the foundation of America's much-heralded vaccine campaigns. It will shake people's faith in vaccinations and—even more importantly—in the system that's developed, recommended, and pushed them on our children in radically increasing numbers.

Howard intends for our meeting to be a heads-up so I can prepare to report the story when the decision is announced. It will be one of the biggest news stories of our time, and it's important to get it right. After the meeting, back at my office, I alert my bosses that a decision on the vaccine-autism cases is imminent, and I have reason to believe it will be Very Big.

Days pass, and there's no announcement. I'm not sure why the delay. Then, finally, the news crosses the wire services. The vaccine court has *rejected* all of the vaccine-autism claims. Denied them. Kaput. *What on earth had happened between my briefing from Howard—and now?*

If I were covering the same story today, I'd have some idea of how to report on the contradiction between what I knew had been decided—and what was announced. I should have immediately started my detective work to report on the discrepancy. Who and what were responsible? But back then, I wasn't as knowledgeable about the secret machinations and how to deal with them. I was just dumbstruck. So the story was covered on the *CBS Evening News* the same way it was everywhere else. There's no vaccine-autism link. The vaccine court says so. Case closed.

It would be several years before I'd develop a reasoned hypothesis as to what had actually happened. One of the autism test cases that would have been "founded," or proven if not for the last-minute reversal was that of Hannah Poling.

Hannah's Precedent

Hannah was described as normal, happy, and precocious in her first eighteen months. Then, in July 2000, she was vaccinated against nine diseases in one doctor's visit: measles, mumps, rubella, polio, varicella, diphtheria, pertussis, tetanus, and haemophilus influenza. Afterward, her health declined rapidly. She developed high fevers, stopped eating, didn't respond when spoken to, began showing signs of autism, and began having screaming fits. In 2002, Hannah's parents filed an autism claim in vaccine court. It became one of the test cases in the Omnibus proceedings. It was a strong case. Hannah's father was a Johns Hopkins neurologist. Her mother, a nurse and lawyer. It would be hard for the government and vaccine industry to convince the media that *these* parents were "nutty, anti-vaccine kooks."

But before the historic precedent of Hannah's autism case could be announced, somebody powerful got to the court. A decision was made to quietly carve her out from the Omnibus group of test cases and replace it with a weaker case that the court could publicly smack down. The switcheroo was made. The government secretly paid the Polings, then had Hannah's case sealed under confidentiality so that no other parents would know about it. The Polings received more than $1.5 million for Hannah's life care, lost earnings, and pain and suffering for the first year alone, and then $500,000 per year thereafter. They weren't allowed to tell anybody. So as far as the public was concerned, there was no precedent. No vaccine-autism link. There would be no flood of similar cases. No financial disaster for the vaccine industry. Even more insidious: knowing full well that the admission had been made in private, officials like CDC director Dr. Julie Gerberding continued to let the public think the vaccine-autism connection was a debunked myth.

In 2008, fate kicked in, and news of the Poling settlement leaked out to the press. I covered the story. Hannah's parents held a press conference and provided credible accounts of their daughter's descent into illness and autism after vaccination. *Time* summed up the relevance this way: "[T]here's no denying that the court's decision to award damages to the Poling family puts a chink—a question mark—in what had been an unqualified defense of vaccine safety with regard to autism. If Hannah Poling had an underlying condition that made her vulnerable to being harmed by vaccines, it stands to reason that other children might also have such vulnerabilities."

By then, Dr. Gerberding of the CDC had gone on to another job. She was now president of Merck Vaccines. From that perch, she addressed news of the Poling precedent. "The government has made absolutely no statement indicating that vaccines are a cause of autism," she insisted. "This does not represent anything other than a very specific situation and a very sad situation as far as the family of the affected child."

Read about the $1.5 million secret settlement paid for the vaccine-autism injury of Hannah Poling here: https://www.cbsnews.com/news/vaccine-case-an-exception-or-a-precedent/

Read the Poling case documents here: https://www.ageofautism.com/2008/02/full-text-autis.html

4. **The Zimmerman Revelations:** A decade after the Poling autism news conference, the government's top expert medical witness, Dr. Andrew Zimmerman, came forward with his own remarkable account of what had actually gone on behind closed doors. He told his full story in 2018 in a sworn statement. Robert F. Kennedy Jr., a lawyer and advocate who was helping fight on the side of vaccine-injured children in court, was instrumental in convincing Dr. Zimmerman to document his incredible claim all these years later in an affidavit. "This was one of the most consequential frauds, arguably, in human history," Kennedy tells me in an interview about the scandal for my television program.

   According to Dr. Zimmerman, on June 15, 2007, he took aside two Department of Justice (DOJ) lawyers for whom he worked defending vaccines in vaccine court. As I mentioned earlier, DOJ lawyers, paid by us, take the side of the pharmaceutical industry against patients claiming vaccine injuries in court. On this day, Dr. Zimmerman told the DOJ lawyers something they likely never expected to hear: that "vaccinations could cause autism," after all, in some cases. In fact, he told them that he'd personally treated such patients. This was akin to an accused killer's best friend admitting his buddy committed murder. The government's own star witness was conceding the vaccine-autism theory was true!

   "I explained that in a subset of children, vaccine-induced

fever and immune stimulation did cause regressive brain disease with features of autism spectrum disorder," Dr. Zimmerman stated in his affidavit. He says it was a Friday when he delivered the unwelcome news to the DOJ lawyers outside the courtroom. Then, over the weekend, the DOJ attorneys called and informed Dr. Zimmerman that his services as their expert witness in court were longer needed. As if that weren't bad enough, Dr. Zimmerman alleges that days later, the DOJ lawyers went on to misrepresent his opinion in court to continue debunking autism claims. Records show that on June 18, 2007, after the DOJ had in essence fired Dr. Zimmerman, one of the DOJ attorneys falsely told the vaccine court, "We know [Dr. Zimmerman's] views on the issue. . . . There is no scientific basis for a connection" between vaccines and autism. A decade later, Dr. Zimmerman called that DOJ statement "highly misleading."

In 2018, when I tried to ask the DOJ about Dr. Zimmerman's allegations, which could merit serious sanctions against the DOJ lawyers, the agency didn't return my calls. Nobody was ever held accountable. To this day, public health officials continue to actively mislead.

As of 2023, the autism epidemic had grown so ubiquitous that the CDC admitted an alarming 1 in 36 eight-year-olds have been diagnosed with the disorder. The true rate is significantly higher, since it's understood that many cases go undiagnosed, and the rate goes up as you look at children under age eight. Can you think of another disorder that went from almost unheard-of to striking 1 in 36 American kids in the span of a generation? Wouldn't it be reasonable to declare a public health emergency and call for urgent study to identify what we're doing to our children that could be to blame? Isn't it one of the biggest failings of our well-funded public health structure that we haven't?

Instead, autism is normalized and billions of dollars are made treating sick children. Autism studies get published and publicized as long as they point to factors like genetic predispositions or age of the father. Our top experts act not the least bit curious to solve the puzzle. They say they can't explain the sudden epidemic—but insist the one thing they've admitted to privately cannot possibly be a factor.

In a Q and A on the CDC website, one question asks: "Do vaccines cause autism spectrum disorder (ASD)?" CDC's answer: "To date, the studies continue to show that vaccines are not associated with ASD."

Read Dr. Zimmerman's full affidavit at this link: https://sharylattkisson.com/2019/01/dr-andrew-zimmermans-full-affidavit-on-alleged-link-between-vaccines-and-autism-that-u-s-govt-covered-up/

5. **Switcheroo, Part Two**: In the middle of covering stories about vaccine controversies in the early 2000s, I keep asking the CDC for interviews. One fateful day, agency officials agree to an on-camera interview with me for the *CBS Evening News*.

   I fly from Washington, DC, to the CDC headquarters in Atlanta. The agency's public relations official greets me, stations me in a waiting room, and preps me for what I'm about to be told in the interview. "We'll be making a big announcement," he says. "We'll be recommending that thimerosal not be used in flu shots for pregnant women."

   I hadn't expected the agency to acknowledge *any* potential risk to thimerosal. CDC officials had long fended off all suggestions that the mercury preservative in vaccines had any downside. The fact that the CDC is now willing to open the door, even a crack, to health concerns about thimerosal would be major news.

   As the CDC press person chats me up, he tells me that he personally still thinks there's really nothing dangerous about thimerosal, but adds that it's easy enough for pregnant women to ask their doctor for a thimerosal-free version of flu shots. He uses the word "nutty" and other slurs to describe parents who are concerned, and says their activism has pretty much forced the agency's hand.

   As I wait to be escorted into the CDC studio where the interview will take place, there's an unexplained delay. The appointed time for our interview passes, and the clock ticks on. Time is of the essence when I'm producing a story for that night's *Evening News*. I wonder what's going on.

At last, I'm escorted to the interview room. I sit down across from the CDC official and ask about the potential risks of mercury, or thimerosal, in vaccines. I wait to hear the CDC announce the new recommendation that pregnant women should avoid it. But to my surprise, the CDC official tells me there is no issue! *Thimerosal is safe. There's nothing to fear.* Thoughts spin around inside my head. After telling me one thing privately, they've decided to toe the industry line publicly. I re-ask the question. I get the same answer.

I have no way to know what happened behind closed doors prior to the interview, but I sense that a great debate has taken place. That our most powerful political figures and drug industry moguls had reached down at the very last minute and stopped the CDC from making its planned statement. Somebody doesn't want the door to be opened even a crack. *Who are the people pulling strings? Where do they hold their secret talks? Who has that kind of power?*

6. **Vaccine-Autism Link "Worthy of Investigation":** In 2008, the medical establishment was busy trying to put to rest controversy over vaccine-autism links. But the former head of the National Institutes of Health (NIH), Dr. Bernadine Healy, bravely told me in an on-camera interview that it very much remained an open question. She said her colleagues at the Institute of Medicine were willfully turning a blind eye to research that could identify children who are "vulnerable" to vaccine injury.

   The interview, like much of my work on medical scandals, was somehow wiped from nearly every corner of the Internet, but I was able to repost a copy on my Rumble page.

   You can watch the interview with Dr. Healy here: https://rumble.com/v23b9tu-dr.-bernadine-healy-interview-with-sharyl-attkisson-on-vaccine-autism-link.html

7. **What's in a Name?** In covering vaccine stories for CBS News, I discovered a telling trend. In vaccine court, when autism is

alleged, the cases are lost. But when the plaintiffs avoid using the word "autism" and instead highlight "encephalopathy" or general brain damage, the cases are often won. That makes no logical sense to families who say that autism is nothing more than a subset of encephalopathy. But as a result of the wholesale rejection of autism claims, some are circulating these words of advice: use the word "encephalopathy" in your vaccine court claim, and you're more likely to win. Call the same injury "autism," and you're sure to lose.

"I purposely avoided mentioning 'autism' in the claim," says the attorney for a child diagnosed with brain damage and autism after her DTaP vaccination at eighteen months. The lawsuit alleged only "encephalopathy," general brain dysfunction. "Using (the child's) autism diagnosis would have dragged out the lawsuit for years," observes the family's attorney. "The point wasn't to try to win the autism debate. It was to get this family the compensation they need to take care of their injured child." They promptly won a significant award.

The case of poor Michelle Cedillo couldn't have turned out more differently because her case did utter the dreaded "autism" word. It was one of the "test cases" in the Omnibus Autism Proceeding we've discussed. Michelle's attorney argued that an MMR shot on December 20, 1995, directly caused her severe autism. But the vaccine court was unequivocal in smacking down the claim in 2009, saying there was no credible proof that vaccines caused her autism. Michelle's diagnosis includes severe encephalopathy. Her mother, Theresa Cedillo, says they might have won their claim if they'd simply called Michelle's brain injury "encephalopathy" and left out the autism part of the diagnosis. Mrs. Cedillo told me she doesn't regret her daughter being a landmark case in the Omnibus proceedings, even though they lost. But for future families, she advises, "If you want to be compensated, I would say stay away from the 'autism' word."

Some years earlier, I'd gotten an inside tip that the vaccine court had been paying autism claims for years under the guise of "encephalopathy." With that in mind, I filed a Freedom of Information Act request to learn just how many brain damage cases had been paid, and what subset of those were actually autism. To my surprise, the government provided some information without much delay. According to the Department of Health and Human Services' Health Resources and Services Administration (HRSA), by May 2010 more than half of vaccine injury awards were given for brain injuries. Here's how the numbers broke down:

Number of Brain Injury Cases Compensated in Federal Vaccine Court

639	Encephalitis or Encephalopathy
656	Seizure Disorders
1,295	Total

But the government told me it could not answer the crucial question: How many of those brain damage cases involve children who also ended up with autism, like Hannah? If the court is compensating more than just a few brain-damaged children who were diagnosed with autism as part of the injury, it's important information. The answer could help prove or disprove the vaccine-autism link. Inexplicably, *the government isn't tracking that data.* Government officials told me, "The government has never compensated, nor has it ever been ordered to compensate, any case based on a determination that autism was actually caused by vaccines. We have compensated cases in which children exhibited an encephalopathy, or general brain disease. Encephalopathy may be accompanied by a medical progression of an array of symptoms including autistic behavior, autism, or seizures." I read this to mean they acknowledge that vaccines can cause brain damage, including autism. Yet the vaccine court and government seem to claim some sort of technical loophole to deny claims when they specifically mention an autism diagnosis.

When I asked government officials at the time why they weren't examining or tracking the rate of autism among brain-damaged vaccine victims, including the thousands who brought cases to vaccine court, they told me, "The Court allowed the filing of 'shortform' petitions, but without medical records. As a result, a very small number of the pending 5,000 claims have medical records, making it impossible for us to review and compare commonalities, patterns, or any general trends among all of the petitioners. Over time, we may learn more about patterns of pre-existing conditions and the role vaccines play, if any, in their progression. As we have done in the past, the [Vaccine Injury Compensation Program] medical staff will look at the court findings and any new scientific information, and may publish scientific articles as appropriate." This explanation is seriously flawed. The vaccine court has a self-selected population of thousands of ready and willing study subjects just waiting to be examined. To look at this wealth of data and let it go unmined makes no logical sense unless you don't really want to find answers. For every vaccine brain injury claim filed, the government need only contact the family for more details, or get access to medical records to look for autism diagnoses among them, and identify patterns that might show why those particular children got hurt.

The fact that they aren't trying to answer the important questions speaks volumes.

Read more on the topic here: https://www.cbsnews.com/news/vaccines-autism-and-brain-damage-whats-in-a-name/

8. **What's in a Name? Part Two:** In 2001, the government addressed escalating worries about the controversial mercury-laden preservative "thimerosal" and its possible links to autism by urging its removal from most vaccines. At the same time, the government still insisted thimerosal was perfectly safe. It seemed to be an effort to split the baby.

   While thimerosal was greatly reduced in vaccines, many public health officials and resources seemed to purposely try to mislead people into believing it was simply no longer in any vaccines their children might get from 2001 on. But that's not true at all.

Today, a large heading on a CDC web page reads, "Thimerosal was taken out of childhood vaccines in the United States in 2001." The thimerosal information page on the website of Children's Hospital of Philadelphia, approved by none other than Dr. Offit, states "Thimerosal was removed from vaccines after an amendment to the Food and Drug Administration (FDA) Modernization Act was signed into law on Nov. 21, 1997." These statements should receive five Pinocchios from any honest fact-checking organization.

For example, the long list of shots that still contained thimerosal in 2004, according to an FDA chart published at the time, include Tripedia's DTaP vaccine, some Td shots, one formulation of DTaP-Hib, a pediatric HepB vaccine, DT shots, Tetanus Toxoid shots, and many flu shots—all of these listed by the government as "routinely recommended pediatric vaccines" or "recommended for some children." Thimerosal also remained in the Hepatitis A-Hepatitis B dual shot at the time, as well as the meningococcal vaccine approved for children as young as age two, and a Japanese Encephalitis shot sometimes given to youngsters age one and up.

It's difficult to square those facts with the public messaging. When I was assigned to cover vaccine safety issues during this time period at CBS, one managing producer had been lobbied by and held meetings with vaccine industry advocates who sought to influence the network's reporting on growing vaccine controversies. One day while this producer reviewed one of my reports prior to air, she parroted the industry-government line declaring, "Thimerosal isn't in any vaccines, anymore." I knew she was wrong and was taken aback by the idea that a news person at the network level would accept the word of a special interest without checking. I was able to disprove her assumption by showing her a government produced chart of thimerosal-containing vaccines.

Sometimes the presence of thimerosal in vaccines is brushed off as just a "trace" without the acknowledgment that even the tiniest amounts may prove harmful to some children and can accumulate with each shot, according to some scientists. You might also see wordsmithing that states a particular vaccine "no longer contains thimerosal *as a preservative* [emphasis added]." Most would infer that to mean the vaccine is thimerosal-free. Yet it might contain thimerosal left over *from the manufacturing process.* Just not as a preservative. If your child has a peanut allergy, would it be okay for products that may contain small amounts of peanut dust to be labeled in a way that implies there isn't any?

As of this writing in 2024, CDC's difficult to find chart of thimerosal-containing flu shots includes Aflura, given to kids as young as age five; Fluvirin, approved for kids as young as age four; and FluLaval Quadrivalent and Fluzone Quadrivalent, both approved for babies as young as six months old. The TDVAX vaccine against diphtheria and tetanus for children as young as age seven also contains thimerosal, according to the chart. In March 2024, the manufacturer announced it was discontinuing that shot, but remaining supplies were still being used up.

When pressed, public health officials are fond of saying that there are thimerosal-free formulations of any shot your child might get. But it's fair to think the vast majority of parents don't know to ask for such a thing. And a cursory search online reveals that not every provider offers thimerosal-free versions.

Tracking down the truth about thimerosal amid a sea of propaganda shows how difficult it can be to find unbiased, straight facts on important health controversies. Government and establishment medicine websites generally promote a singular view without fairly acknowledging opposing studies or research except to dismiss them. Everything they publish is designed to persuade you that thimerosal is safe, while they paradoxically devote quite a bit of ink to explaining how they've been working for decades to eliminate it from all vaccines anyway.

The public and many in media tend to think of the CDC as an unbiased expert resource for health advice and information. That's not what the agency is. The CDC and some health officials serving in other public bodies see their mandate as persuading us to believe certain things, behave a certain way, and take certain vaccines or other medicine for the supposed good of society. That means they typically cherry-pick through the available data and research, promoting what they wish and discrediting or ignoring the rest. What's worse, critics have built a strong case over the decades that these self-appointed, unelected guardians of our health have sometimes been hopelessly captured by big money interests that can benefit from the actions we're urged to take.

Here's a list of how much thimerosal is in which flu shots (search the page for "thimerosal" to find chart: https://www.cdc.gov/mmwr/volumes/72/rr/rr7202a1.htm

9. **"Autism" Listed on Vaccine Warning Label**: The DTaP shot, or diphtheria and tetanus toxoids and acellular pertussis vaccine known as Tripedia, listed "autism" as a reported adverse event on its package insert. It was right there, plain and simple, in black and white, on the label, at the very time when vaccine interests were insisting that it was a "debunked conspiracy theory." When vaccine safety advocates discovered the autism mention and began publicizing it, the Internet began to fill with "fact checks" to undermine it. Some of the fact checkers falsely claimed "autism" simply wasn't on any vaccine labels. Others said the listing of "autism" under adverse events should be disregarded as a big mistake—that it was included in error only because some misguided parents reported it. If you understand even a little about the review process drugmakers go through to add a single word to their warnings, you'd know how silly that claim is. Interestingly, the Tripedia shot was discontinued in 2013. Not long ago, you could still easily find the Tripedia package insert and the "autism" mention on the FDA's website, but it now seems to have been wiped. You can find it archived at the link below.

   Read the vaccine package insert mentioning "autism" here: https://wayback.archive-it.org/7993/20170112211659/http://www.fda.gov/downloads/BiologicsBloodVaccines/Vaccines/ApprovedProducts/UCM101580.pdf

10. **CDC "Cover-up"**: In 2014, a CDC senior scientist, Dr. William Thompson, confessed that he and his CDC colleagues had conspired to cover up a vaccine-autism link in a study of black boys. He further claimed that he and his CDC study partners "scheduled a meeting to destroy documents related to the study." Dr. Thompson testified that they "all met and brought a big garbage can into the meeting room and reviewed and went through all the hardcopy documents that we had thought we should discard, and put them into a huge garbage can." *They literally trashed study documents.* Dr. Thompson became a whistleblower, hired a lawyer, saved copies of the original data in a safe, and turned it over to the only member of Congress willing to look at vaccine safety scandals, Representative Bill Posey, a Republican from Florida. Posey could not convince anyone in congressional leadership to investigate the matter and was mercilessly attacked by the media and other vaccine industry interests when he spoke about Dr. Thompson's allegations. The other CDC scientists who were implicated denied doing anything improper. And the CDC has defended the controversial study as published.

    Read more about Dr. Thompson's allegations here: https://sharylattkisson.com/2021/01/cdc-scientist-we-scheduled-meeting-to-destroy-vaccine-autism-study-documents/

    Watch Representative Bill Posey's statement on Dr. Thompson's testimony: https://www.c-span.org/video/?c4554834/user-clip-rep-bill-posey-cdc-whistleblower

11. **CDC Official Says Vaccine-Autism Link "Possible"**: While the CDC consistently misrepresents studies on vaccines and autism, implying there's no possible link, Dr. Frank DeStefano, director of the CDC Immunization Safety Office, made an important admission in a 2014 telephone interview with me. Dr. DeStefano was one of Dr. Thompson's colleagues who allegedly took part in trashing data from the study about autism in black boys. Anyway, Dr. DeStefano acknowledged to me that vaccines may, rarely, trigger autism in susceptible children. This was quite an admission, and came only after I phrased the question in a way

that he could not easily dodge. When I persisted with follow-up questions, Dr. DeStefano suggested that somebody should conduct studies on the topic. Did the National Institutes of Health step forward and carve out a slice of their research billions to answer this important question? *Nah.*

Read more, and listen to the interview with Dr. DeStefano at this link: https://sharylattkisson.com/2018/12/cdc-possibility-that-vaccines-rarely-trigger-autism/

12. **No Flu Shot Benefits for Elderly:** During a flu shot shortage in 2004, the medical establishment had used such frightening scare tactics to sell vaccines that thousands of elderly Americans stood in long pharmacy lines trying to score a shot. There were reports of some passing out or even dying while waiting for their jab. The emergency medicine chair at Emory University issued an ominous warning. He said, "The combination of the vaccine shortage, more than 80 million Americans at high risk of flu complications, and a nationwide emergency department crowding crisis means . . . the prospect of the 'perfect storm'—a surge of critically ill flu patients and no resources to care for them."

    No such disaster materialized. But given the government's hard-sell tactics, it might surprise you to learn that, as government scientists later explained to me, *no* definitive *study had ever proven that flu shots even work*! It turns out that over the decades, our top experts were doing something very unscientific—the polar opposite of "following the science." They were *rejecting* what the science really showed about flu shots.

    Honest, top government scientists who later ended up revealing the truth explained to me that the best studies had long indicated flu shots don't work. But the medical establishment agreed, as a group, that "the best studies"—had to be wrong. *Flu shots* must *work, they told themselves; they* had *to work!* So they became conveniently selective about the flu vaccine studies they would cite. They tended to rely on less definitive but positive-sounding reports. And they would twist data from negative studies to try to make the results seem positive. It's hard to imagine a more serious violation of sound scientific practices.

Faced with the persistent and inconvenient facts, and ever determined to shape the science to fit its desires, the government launched a major study designed to hopefully show, once and for all, that flu shots *do* work. The vaccine industry surely favored that approach. The esteemed government researchers assigned to conduct the study, who later spoke with me, admitted they began with a bias in favor of flu shots. Yet they said that no matter how they crunched the data or tweaked the stats, they ended up with the same awful truth: as more and more older people got flu shots, more of them died.

Specifically, they concluded, "We could not correlate increasing vaccination coverage after 1980 with declining mortality [death] rates in any age group . . . we conclude that observational studies substantially overestimate vaccination benefit." Studies in Italy and other countries found the same problem with flu shots.

I was pretty shocked too when I learned about all of this in 2006 after the government study was published. (In today's scientific environment, I don't think the study would have seen the light of day since it produced an undesired answer to the question.) I was able to talk to some of the study scientists themselves to get background and context for my reporting. They confessed that they were stunned by their own findings. They knew the implications were dire.

This blockbuster news about the ineffectiveness of flu shots should have made global headlines, led to an international reexamination of flu vaccine policies, and prompted a serious analysis of how our top officials could have gotten things so wrong. But none of that happened. US government officials would not allow the scientists to do on-camera interviews with me, and the CDC decided to continue recommending flu shots for the elderly.

I asked a top government vaccine source at the time why they weren't going to back off flu shots in light of their own study. He told me, "We can't take flu shots away from the elderly after we've spent so many years convincing them they'll die without them." He then explained that the government would "have to come up with a new strategy." He told me that would likely mean they'd start recommending flu shots for children for the first time so that they don't carry the flu to the elderly. The official told me, "The hard part will be convincing parents to give flu shots to their kids when it's not for their child's benefit." Indeed, the following year, the CDC began recommending flu shots for children, insisting it was for their own good. And the government still recommends annual flu shots for the elderly too, spending hundreds of millions of tax dollars to pay for them.

The less helpful the vaccine proved to be, the more people were urged to get it.

Read the full flu shot study article here: https://jamanetwork.com/journals/jamainternalmedicine/fullarticle/486407

Read my full report on the study here: https://sharylattkisson.com/2022/11/govt-study-finds-flu-shots-not-effective-in-elderly-after-all/

13. **Polio from Polio Vaccine:** Though parents were not routinely told this alarming fact, the oral version of the polio vaccine, sometimes referred to as "sugar water," carries a slight risk of giving the child polio. For years this version was given to American children even though there was a safer injectable version that had no chance of transmitting polio. The oral version was finally pulled off the US market in 2000 but is still used in other countries. The last cases of polio in the United States were all caused by the oral vaccine. The first outbreak in the US in years, in New York State in 2022, was from an oral polio vaccine strain, according to the CDC. The oral vaccine strain also carries a risk of morphing into a more dangerous version. The CDC responded to news of the 2022 outbreak by announcing that to try to curb it, the agency is considering restarting controversial oral polio vaccinations in the US—the ones that can cause polio.

    Read more on the "vaccine-derived polio outbreak" in New

York at this CDC link: https://www.cdc.gov/media/releases/2022/s0913-polio.html

14. **Swine Flu Exaggerated:** In 2009, based on two insider tips, I began investigating America's supposed "swine flu epidemic" and learned there was almost no actual swine flu circulating in the US. The CDC was fabricating the emergency.

    I made the discovery based on an examination of lab results from all fifty states, which I had to obtain state by state because the CDC refused to promptly turn over the data it had from the states even though it was required to do so under Freedom of Information law. The state data was important because it contained lab test results of ill patients most likely to have swine flu, based on their symptoms and risk factors. In other words, these "most likely" specimens should have tested nearly 100 percent positive for swine flu if the CDC were correct that we were suffering a true emergency. But I found a far different reality. There was very little swine flu. In Florida, out of all the specimens believed to be swine flu, just 17 percent actually turned out to be swine flu. In Georgia and California, only 2 percent of supposed swine flu specimens were swine flu. And in Alaska, only a scant 1 percent of the samples deemed most likely to be swine flu—tested positive for swine flu.

    Despite the fact that a swine flu epidemic didn't exist, the government pressed forward with a hastily developed "emergency" swine flu vaccine that ended up causing numerous injuries. A few of the vaccine-injured patients were ultimately compensated in vaccine court.

    Read my swine flu investigation here: https://www.cbsnews.com/news/swine-flu-cases-overestimated/

15. **Vaccine Ad Loopholes:** If you're young, you don't remember a time before drug ads flooded the television airwaves and Internet. As I've mentioned, ads for prescription drugs used to be banned on TV and are still prohibited in almost every other nation on the globe. The long-standing and prevailing belief is that only trained medical professionals should be determining when to prescribe which drugs to whom.

But drug ads were permitted in the US after a fierce lobbying campaign by the pharmaceutical industry. Now companies spend billions of dollars a year on this advertising, creating the mutually beneficial but ethically conflicted dynamic between news, media, and Big Pharma.

As a supposed safety measure, drug ads are required to clearly disclose some risks. Yet when it comes to commercials for vaccines, you'll often see no mention of risks at all! The advertisements leave the false impression that there's simply no possible downside. I decided to look into how vaccines managed to escape FDA rules that require risks to be disclosed.

It turns out vaccine makers are exploiting a type of loophole, and the FDA allows it. Vaccine commercials fall under a conveniently created category the FDA calls "disease awareness ads." In short, as long as the ads don't mention a specific brand, such as Pfizer or Merck, they're considered "educational" and don't have to tell consumers about any risks. The justification given is that these ads aren't really pushing a particular medicine. They're just creating awareness about medical conditions. But critics say it's a dangerous practice that allows drug companies to promote vaccines in a slanted fashion.

For example, a GlaxoSmithKline TV ad for adult whooping cough vaccine (that doesn't mention GlaxoSmithKline) doesn't warn about the vaccine's links to paralysis and possible heart problems. It also doesn't mention that 4 percent of patients who get the adult whooping cough vaccine report having serious adverse events within six months.

Merck's disease awareness ads for HPV cervical cancer vaccine prey on parental guilt. And because they don't say the word "Merck," there's no mention required of the vaccine's risk of blood clots, seizure, appendicitis, and paralysis. The ads falsely make it seem as though there's only an upside to getting the vaccine—and that you're a bad parent if you don't take your kid to the doctor to get it.

In early 2019, I asked the FDA about "disease awareness"

advertising and the loopholes that allow risks to be hidden. A spokesman told me he understands the controversy and assured me the FDA was looking into the practice to determine if it "may result in consumers misinterpreting and being confused." However, there was no apparent movement on that front before Covid hit. Naturally, when the Covid vaccines went to market, the media was flooded by "disease awareness" ads that failed to disclose any of the serious safety concerns, exaggerated vaccine effectiveness, disregarded natural immunity, and promoted vaccination for all, even if it could prove life-threatening for some.

Read more about drug ad practices here: https://fullmeasure.news/news/cover-story/the-pill-pitch

That's a lot to absorb! The point is to give you a taste of some startling facts that are little reported but provide important balance to widespread claims in the media. It all makes a strong case for the idea that we cannot accept at face value scientific information from the media, federal agencies, medical journals, government and industry experts, medical professionals, academic researchers, or public health officials.

This is vital background bringing us to the moment when Covid changed all of our lives.

CHAPTER 7

De Niro and Me

It's Tuesday, April 12, 2016. I'm driving to a work appointment when my cell phone chimes. I don't usually pick up numbers I don't recognize. But this call is from "Unidentified" and sometimes that means a member of Congress or high-ranking government official. I decide to pick up. It's a woman's voice.

"Hello, is this Sharyl Attkisson?" the voice asks.

"Who's calling?" I reply.

"This is Grace Hightower De Niro," she says. "Bob would like to talk to you. Are you free to do a call?"

Bob? Linking to the name "De Niro," I figure "Bob" must be "Robert" De Niro. *So—he goes by "Bob"?*

We set up a call for later that morning. Before we hang up, Grace asks what I think of British gastroenterologist Dr. Andrew Wakefield. The answer is complicated. I tell her he's been smeared and destroyed by those at the highest levels who make a living doing that sort of thing on behalf of pharmaceutical interests. I tell her he's also one of the most knowledgeable sources on the topic that I know is about to be discussed with the mention of his name. Grace agrees Dr. Wakefield is a credible authority. She's already done some homework scratching beneath the superficial propaganda. *Interesting.*

A short time later, Grace calls back. Bob can do the call in five minutes . . . sooner than we'd scheduled. We hang up and the third call comes. Now the three of us are conferenced in together. Grace explains she's at dance class with their four-year-old daughter. Bob is speaking from another location.

Grace and Bob tell me that people they've consulted and trust recommended they talk to me to get some unbiased information. They've also been pointed to Robert F. Kennedy Jr., comedian Rob Schneider, and Congressman Bill Posey. All of those people and Dr. Wakefield have one thing I know in common: they've done deep research on the links between childhood vaccines and autism, and the decades-long government, pharmaceutical industry, and media cover-ups.

De Niro himself has already been making international headlines on that very topic. As cofounder of the famous Tribeca film festival in New York, he'd recently announced plans to screen Dr. Wakefield's new documentary, *Vaxxed: From Cover-up to Catastrophe*. The high-profile showing of *Vaxxed* would make for a watershed moment in the movement trying to expose the vaccine-autism connection and cover-up. Now De Niro, one of the biggest celebrities on the planet, was asking questions and considering all sides. And unlike everybody else, it looked like Bob "Are-YOU-Talking-to-ME?" De Niro wouldn't be easy to intimidate, smear, or cancel into oblivion.

Vaxxed documents allegations about the CDC's destruction of data from a 2004 government study linking autism, particularly in black boys, to the measles, mumps, and rubella (MMR) vaccine. The film centers on the senior scientist at CDC who led the study, Dr. William Thompson.

A few years earlier, Dr. Thompson had privately reached out to a biologist named Brian Hooker. Hooker was trying to get the CDC to cough up information as part of his personal investigation. His own son had developed autism after vaccination. In phone conversations over a period of several months, Dr. Thompson confessed to Hooker that he felt guilty about CDC cover-ups of vaccine-autism links, and helped Hooker unravel details. Hooker secretly recorded his calls with Dr. Thompson and eventually made them public. Dr. Thompson went on to hire an attorney and become a whistleblower, providing a full confession about CDC's misdeeds to Congressman Posey—only to make the harsh discovery that Congress at large wasn't interested, and the entire controversy impacting millions of children was taboo on Capitol Hill and in the media. *Vaxxed* chronicled this scandal.

Dr. Wakefield directed *Vaxxed*. For his part, he'd found himself targeted years earlier by the vaccine industry and its surrogates. That was after he had the misfortune in 1998 of conducting a study that unearthed a potential link between MMR vaccines and autism, the very thing Dr. Thompson was attesting to years later. Despite widespread misreporting that claims otherwise, Dr. Wakefield never alleged in his study that vaccines cause autism, and he didn't urge parents to reject vaccination. He simply concluded that, based on his research, the topic needed more study. After all, nobody had officially cracked the code as to what was behind the sudden autism epidemic, and his findings raised questions about a possible vaccine link. Other important scientists signed on to the study when it was published. Back then, more than twenty-five years ago, it was before pharmaceutical interests came to control virtually every aspect of information on the topic. Dr. Wakefield had no way to know, but he'd stumbled onto information so massive and potentially damaging to the multitrillion-dollar vaccine business that industry interests had to neutralize him quickly and permanently. He became a victim in one of the most ruthless and well-orchestrated smears of our time.

De Niro's interest in showing *Vaxxed* at Tribeca had now triggered similar panic among the pharmaceutical industry and its allies. De Niro quickly fell under so much pressure from all directions that on March 27, 2016, about two weeks before our first phone call, he announced he was canceling the *Vaxxed* screening. The documentary world used to be considered the last place where it was safe to examine controversies of all kinds with a type of freedom once encouraged in the media at large. But now vaccine industry interests were working to block the subject from being explored fairly and fully, for fear the public would dare to draw their own conclusions.

In that first phone conversation with Bob and Grace, I don't do much talking. I listen. It becomes clear that the couple are information-seekers. As they speak about why they're researching vaccines and autism, they sound rational and fair. They tell me their now-teenaged son has had autism since his early years, and that Grace has a "searing memory" of his devolution from normal—to deeply affected after vaccination. There was a "marked difference." They liken the whole controversy to that of the tobacco industry, and the link between smoking and lung cancer that was covered up for so long.

Bob tells me that making headway in getting and telling facts on this topic is "like pushing an elephant uphill." He asks me about the CDC's Dr. Thompson and why one of the seemingly biggest stories of our time hasn't been widely told. He also asks me about Dr. Wakefield and, on the other side, vaccine industry advocate Dr. Ian Lipkin, who'd apparently assured De Niro that vaccines couldn't possibly be linked to autism. I give De Niro a very short summary of what I know about the two men.

Dr. Lipkin, director of the Mailman School's Center for Infection and Immunity at Columbia University, had recently signed an op-ed in the *Wall Street Journal* and "worked diligently behind the scenes to persuade organizers" that Dr. Wakefield's documentary had no place at the prestigious Tribeca Festival. Like many public institutions, Columbia University receives generous funding from the vaccine industry and its allies. I think that helps explain why academic and research leaders like Dr. Lipkin might put themselves in the peculiar role of censoring medical and scientific information.

The next thing Bob De Niro says is intriguing. He still wants to hold a screening of *Vaxxed* but outside of Tribeca. Maybe, he says, in the next week and a half or two. He's trying to think of the best way to do it. Perhaps he'll simultaneously screen related films, such as *Trace Amounts*, a documentary examining the role of mercury in the autism epidemic. "Wakefield [of *Vaxxed*] gets so much knee-jerk reaction," Bob observes. "We could run *all* these films that expose the issue. It would be a great idea. Good to let the public know. They need to know."

The idea that Bob has suffered through the Tribeca public relations mess but is still determined to get into the vaccine-autism controversy is remarkable. *They got the Tribeca screening canceled, but they didn't get to De Niro,* I think to myself. I then take the opportunity to ask what really happened at Tribeca. Who intervened and forced his hand to cancel the screening? Bob says he was told by somebody he didn't name that "It was not the sponsors. It was the other filmmakers" who objected to the screening.

Ah, the plot thickens. Bob is made to believe there's overwhelming, grassroots disapproval of *Vaxxed* among filmmakers. Yet whoever made that claim didn't tell him exactly *who* the objecting filmmakers are. "I didn't ask," he goes on to say. "I didn't want to go any further." He then muses, "I don't know if I made a mistake" to pull the film.

I find it unlikely that a bunch of documentarians got together to try to cancel another documentarian. Bob was specifically told it was *not* the sponsors who objected, leading me to believe that's exactly who it was—but that they wanted no fingerprints on their handiwork. Later, when I look up the list of Tribeca Festival sponsors, among them is none other than vaccine advocate and funder Alfred P. Sloan Foundation. *Any form of information that can be bought, co-opted, or slanted—has been,* I remind myself.

The Tribeca controversy aside, I note that Bob has started to get up to speed on vaccine science impressively fast. It took me more than a year of examining research, developing credible sources, and speaking to scientists before I felt ready to do my first story about childhood vaccines and autism on CBS News years ago. Like almost everyone else, I'd been told the link was a "debunked conspiracy theory." I was as shocked as anyone to find an impressive body of research, scientists, and court cases proving otherwise. One of the hardest things to do as a reporter, perhaps as a human, is to admit that something you thought to be true turned out to be wrong—especially when reporting the truth means swimming upstream against the most powerful forces. But over the years, I've become pretty good at doing just that. I've come to invite the opportunity and embrace the intellectual exercise of changing my mind or at least reconsidering what I think I know, when merited.

Bob De Niro gives the impression of an independent thinker. An intellectual with a scientific mind. I can tell he's going through his own awakening of sorts, even if he's not quite there yet. Like me (years ago), he's stunned to find credible studies and scientists giving full validity to the vaccine-autism link after being told they don't exist. He can't fathom why the government would cover up or lie.

Grace leaves the call, but Bob stays on a bit longer. He tells me he's interested in hearing the pros and cons, and in inviting both sides for a fair discussion at the screening he envisions holding. He thinks people will embrace an evenhanded approach, just hearing the facts and letting those who support them and those who dispute them speak openly. *Yeah, that was once me too.* It's a noble idea, but naive. He doesn't yet understand that vaccine industry interests will do anything to prevent a fair, evenhanded discussion of the facts. That's the thing that they fear the most. But I keep my mouth shut on this point. Maybe Robert De Niro can pull off something nobody else has been able to do. Maybe he, uniquely, has the power to do it.

Later that afternoon, the phone rings again. It's De Niro's assistant. Bob wants to talk again.

"Sharyl, I'm going to be on the *Today* show with Matt Lauer in the morning about the [Tribeca film] festival," Bob tells me. "They might ask me a question on vaccines."

There's no "might" about it. They *will* ask. Someone likely pulled strings and got him invited on the NBC program for that very reason: to force a public mea culpa, where De Niro contritely tells the world he was mistaken to invite a debunked documentary by a discredited doctor to be screened at the esteemed Tribeca Festival.

"They'll ask if I'm anti-vaccine," he speculates. "I'll say, 'No, I'm pro-vaccine.'"

It won't matter, I think to myself.

There's no bigger propaganda feat of my time than the popularization of the invented term "anti-vaccine" in the early 2000s. The propagandists apply it liberally to scientists, journalists, parents—*anybody* who researches vaccine safety or reports on it with facts that come down on the "wrong" side. It's not just used to label controversial assertions. Indisputable facts about vaccine safety can be called anti-vaccine just because of who said them. Even people injured by vaccines are called "anti-vaccine." When examined academically, the term makes no sense. Obviously, parents can't be dismissed as "anti-vaccine" for having a child injured by vaccines—they vaccinated their child in the first place! Scientists aren't "anti-vaccine" simply because they discovered adverse events connected to vaccination. I'm certainly not anti-vaccine for reporting on vaccine safety issues. My daughter and I have had all the recommended vaccines—and then some.

But the phrase is uncannily effective. There's something about falsely tagging someone as "anti-vaccine" that seems to have a magical effect on others. Logic is tossed aside, reporters reject normal fact-finding practices, minds are tightly shut. Nobody bothers to consider the existence of solid studies exposing vaccine side effects. Instead, they robotically insist any negative facts about vaccination cannot possibly be true. Only those putting out pro-vaccine narratives with cultlike fervor are to be heard or believed. The public and media tsk-tsk and nod their heads knowingly about the anti-vaccine "nuts" or "crazies."

I was startled when I first heard those slurs applied to parents of vaccine-injured children about fifteen years earlier. I was assigned to cover the subject for CBS News, and the name-caller was a CDC PR man. Back then it was shocking to hear a government official call concerned parents "nutty." Today I take for granted that some of the worst propaganda comes from our public agencies.

For Bob De Niro, I intuitively know that the key to surviving the *Today* interview with his reputation intact lies with him remaining resolute. The mistake I've seen too many make is getting bullied into backing down from asking rational questions about vaccines. They apologize. Recant. Promise to keep their mouths shut. They hope to save their careers or whatever is being threatened. But it's a trap. Their apologies only make things worse. Now the propagandists can seize upon the moment to lump the apologetic person in with a long list of "discredited" scientists, celebrities, and other "anti-vaxxers." They're looking for weakness. *Show any chink in the armor, and they'll eat you alive.*

The next morning, Wednesday, April 13, 2016, I turn on the *Today* show. As I'd predicted, the vaccine kerfuffle isn't just one question to

De Niro; it's an entire discussion, which takes place with NBC reporter Willie Geist. Seated beside De Niro is Jane Rosenthal, cofounder of the Tribeca Festival.

During the interview, De Niro stands firm and advises viewers to see *Vaxxed*. In a weird dynamic that surely wasn't choreographed by the PR consultants, De Niro and Rosenthal go back and forth, seemingly on different pages. "Read the reviews," Rosenthal recommends. De Niro presents a warm and relaxed demeanor. In a measured tone, he insists that there are questions about the vaccine-autism link, and that the media should report on them.

"I think the movie is something that people should see," De Niro says on the *Today* show. "There was a backlash, which I haven't fully explored, and I will, but I didn't want it to start affecting the festival. But definitely, there's something to that movie. There is another movie called *Trace Amounts*. There's a lot of information about things that are happening with the CDC, the pharmaceutical companies. There's a lot of things that are not said."

"Was it a mistake to pull it?" De Niro goes on to ask himself during the interview. "I don't know. But people should see the film." As for why he withdrew *Vaxxed* from Tribeca, De Niro explains he was "on a shoot" when the whole controversy arose and didn't want it to impact the film festival "in a way he couldn't see." Rosenthal repeats what De Niro had told me on the phone the day before. She says it was "filmmakers, not sponsors" who had complained about the prospect of showing *Vaxxed* at the festival. De Niro counters Rosenthal by saying he finds it "hard to believe" that Tribeca filmmakers objected. He adds that he intends to find out who the true objecters are. *Good for you*, I think.

That night, at his request, I email Bob a "pros and cons" note where I succinctly lay out some of the research and various arguments for and against the vaccine-autism link. The next day, he calls again. He's done more research, personally speaking to some of the key figures who have worked so hard to smear Dr. Wakefield. He's adding things up. He wants to know what I think of three aggressive vaccine defenders: a writer named Brian Deer, Dr. Paul Offit, and—again—Dr. Lipkin. He'd already spoken to Dr. Lipkin, who implied that Dr. Wakefield's research potentially linking MMR vaccines to autism was discredited.

But De Niro is listening to his cognitive dissonance. He's behaving like the right kind of investigator, taking an open-minded and factual approach. He still doesn't realize that the screening he wants to hold, with the promise of "both sides" being heard, will never be embraced by vaccine cultists and propagandists. There's no way they will legitimize his conference.

I think about an incident years before when the head of the special federal vaccine court, Chief Special Master Gary Golkiewicz, asked me to moderate a discussion on vaccines and autism at an upcoming conference. I was an investigative reporter at CBS News. The conference would be attended by lawyers who litigate injury claims in vaccine court. Merck vaccine inventor Dr. Offit would be on the side of claiming no links between vaccines and autism; the former head of the National Institutes of Health, Dr. Bernadine Healy, would discuss science to the contrary. The conference was arranged, advertised, and set. However, shortly before the date arrived, I got an apologetic email and phone call from Golkiewicz. He had been ordered to cancel me and Dr. Healy. Only the vaccine industry's Dr. Offit would be allowed to present at the meeting. There would be no dual-sided discussion or debate, and so nothing for me to moderate. Golkiewicz told me he was terribly embarrassed and regretful to have to make the last-minute cancellation. *Who has the power to instruct the head of the vaccine court to disinvite presenters at his conference?* Golkiewicz tells me, "Let's just say that we on the court do not operate in a political vacuum."

"What does that mean?" I ask. "Is it somebody in Congress?"

He simply repeats himself. "We do not operate in a political vacuum."

In our next telephone conversation, Bob De Niro asks if I'm willing to speak to a public relations guy who's helping him with his idea to arrange a "world-class event" to show various vaccine-autism documentaries and raise public awareness. Bob envisions simultaneous rallies and screenings at locations around the world. The Internet will be utilized, he says, instead of regular TV channels, so that the normal power brokers can't reach in to cancel the plans. *Smart.* I want to hear what De Niro's PR guy has in mind. If it comes to fruition, the affair will be worthy of news coverage.

That afternoon, the PR man calls. He's John Raatz of the Visioneering Group. He tells me he thinks he can turn Bob's idea into a daylong "story-driven, scientist-driven" event with a rational discussion. Maybe, he says, they'll livestream it and make the whole thing into its own worldwide film! The message will be that the public and elected officials should research the topic themselves, answer outstanding questions, and subpoena CDC's Dr. Thompson to testify. But I sense a chink in the armor. Raatz seems nervous about being thought of as associated with Dr. Wakefield, the producer of *Vaxxed.* He says he wants to "distance" from Dr. Wakefield. And, with that, I know. *You may not realize it,* I think to myself, *but the vaccine industry has already won.*

The following day, Friday, April 15, 2016, Bob calls while he's getting ready to tape a show with comedian and talk show host Jimmy Fallon. He tells me that Raatz wants to "throw the baby out with the bathwater." I assume he's referring to Raatz not wanting to associate with *Vaxxed.* I'm not surprised.

Saturday, April 16, Bob calls three more times, chewing on his next steps. His mind seems to be going a million miles an hour. He's trying to think of ways to get CDC whistleblower Dr. Thompson subpoenaed by a reluctant Congress so that Dr. Thompson can tell his story "in a way that cannot be swept under the rug." Republicans are in charge of Congress at the time, and I explain to Bob that their leadership will never allow such a hearing to be held. Frankly, it would be no different if Democrats were in charge. Both political parties are hopelessly conflicted at the highest levels. They get too much money from the vaccine industry.

Bob wants to know which Democrat candidate for president, Bernie Sanders or Hillary Clinton, would be better in allowing the truth to come to light. I speculate it would be Bernie or Donald Trump, the Republican candidate. That's because I sense that neither of those men is reliant upon pharmaceutical industry money. In fact, I tell

Bob, Bernie recently showed some independence by questioning the FDA on some recent controversies. And Trump appears to be open to outsider opinions.

Bob calls back that evening and says maybe somebody like Hollywood producer Harvey Weinstein could do a documentary that's "credible and definitive" on the issue and can't be controversialized the way *Vaxxed* has been. (This is two years before the Weinstein sex scandal was exposed in the media.) Bob has also been looking into congressional hearings held fifteen years earlier by Republican Dan Burton of Indiana before those types of hearings were secretly forbidden by unseen power brokers once and for all. Burton has a grandchild who he says "regressed into autism overnight after his vaccinations." Bob observes that Burton doesn't seem like a "nut" after all, though he was widely portrayed that way in the media after he took on the vaccine-autism issue.

Sunday, April 17, the phone rings around noon. Bob and Grace are on speakerphone together. Their son is there too. Today they're asking me what I know about one of my colleagues, investigative journalist David Kirby. In 2005, Kirby wrote *Evidence of Harm: Mercury in Vaccines and the Autism Epidemic*. That was when a good investigative book on vaccines and autism could still win journalism awards. Kirby was a solid authority on the subject. But the whole topic was later shunned by journalism groups and awards, which became co-opted by pharmaceutical interests.

Bob and Grace are back to wondering how to get Congress to do its oversight job. Could a petition by parents pressure Congress into investigating? I tell them there have been many petitions. That won't do the trick. Grace, who's black, brings up a civil rights aspect to vaccine policy and injuries. She asks, How are kids being given medicine without the full facts disclosed? Black kids suffer from autism at a disproportionately high rate. The study cover-up at the CDC that Dr. Thompson exposed showed an increased autism risk for vaccinated black boys. I agree that a civil rights angle could be more difficult for public officials to brush off.

Friday, April 22, I'm heading to New York City for a screening of *Vaxxed* in a small, independent theater as part of the Manhattan Film Festival. The showing at the much smaller festival was hastily arranged when *Vaxxed* was booted from Tribeca. Bob calls. He says he wants to attend the screening to show support but has other obligations. Grace is busy too. He mentions that Grace has recently been able to connect with Dr. Wakefield. On another topic, Bob notes that there's a good argument on the side that MMR vaccine can't be the cause of autism: it doesn't contain the problematic mercury ingredient, thimerosal, that some blame. Again, De Niro is systematically going down a list, much as I did years ago. Research. Ask. Prove. Disprove. Narrow down. *Methodical.*

What he doesn't yet realize is there are numerous theories about exactly what causes vaccines to trigger autism in some children. In simple terms, as I've explained, some scientists and advocates blame thimerosal. Others believe it's the live virus component of the MMR vaccine. Or it could be the combination of DTaP vaccine, which has been linked to brain injury, followed closely by the live virus MMR vaccine. Or it could be the use of the fever-reducer acetaminophen after a child's vaccine-induced fever. Or it could be an immune response some children have when their systems are challenged with vaccines. The list goes on. One can find research, vaccine court cases, and scientific studies that support each of these ideas.

We know that vaccines can cause injuries. The discussion should be about what they are, how common they are, who is more likely to experience them, and how to prevent them.

Often, advocates backing one particular theory work at cross purposes. I've watched for years as vaccine industry propagandists work to pit advocates and their theories against one another to sow division, doubt, and confusion.

What I've come to learn is that vaccines can cause or trigger a wide variety of brain damage, including autism, and other injuries. How the trigger works in a particular case depends on the vaccines involved, when and how they're administered, and the child's genetic factors, predispositions, and exposures. Adults, too, can get brain damage, paralysis, immune disorders, and other injuries after vaccination, as clearly documented through vaccinated military troops. But the fact that there isn't one simple mechanism to point to understandably confounds people like Bob when they first begin their deep dive and try to wrap their arms around the enormous controversy.

As the days go by, the global event Bob envisioned to prompt a fair, productive vaccine-autism conversation never takes shape. It's clear to me that he's hitting roadblocks with all the entertainment types he would need to help him pull off such a thing. He's learning firsthand how deep the pharmaceutical tentacles and fear over the mere discussion reach. As for those who aren't compromised but still won't touch the subject—maybe they simply believe the disinformation. Or they know better but can't afford to be viewed as "anti-vaccine." They've seen what happens to those who get targeted.

Still, Bob isn't giving up. A few months after the Tribeca debacle, he calls and tells me he's speaking with director Robby Kenner about doing a documentary on vaccines and autism that would be "beyond criticism." Kenner directed the documentary *Food, Inc.*, which bravely took on corporate farming in America. He would seem to be a fearless choice for Bob's vision of a film taking a hard, truthful look at vaccines. Bob asks if I would be willing to have a conversation with Kenner to give some factual background on the controversy. I agree—but Kenner never calls. A few weeks go by, and I tell Bob I never heard from Kenner. Bob's assistant connects me directly to Kenner via email. Kenner responds by saying that he'll call me the following week, but he never does. Now I'm curious. Is Kenner doing a project with Bob or not? Another month goes by, and Kenner finally emails and sets up a call. He talks and I listen. He speaks at length about his interest in doing a film about the pharmaceutical industry and its history, influence, and control. But one important piece seems to be missing. He never once says the word "vaccine." When he finishes, I ask if he's going to address the vaccine and autism controversy in this project of his. He answers that it's far too ticklish. Best to avoid, he says.

So there you have it. It doesn't seem to matter who Bob contacts. Nobody will do his vaccine project.

On February 15, 2017, I finally get to meet Robert De Niro in person after all of our phone calls and email exchanges. He's appearing in Washington, DC, where I work, at a press conference held by Robert F. Kennedy Jr. on the topic of vaccine safety. Kennedy has recently been contacted by the newly elected president Donald Trump about forming an independent scientific committee on vaccine safety. Kennedy says Trump told him that he knows the vaccine industry is going to "cause an uproar" over it, but Trump says, "I'm not going to back down."

At the Washington, DC, news conference, Bob is reserved and thoughtful. Afterward, I interview both him and Kennedy for a report on my TV program *Full Measure*. Bob tells viewers what got him interested in the vaccine-autism issue.

"I mean, I never was really that aware other than that my own son was on the [autism] spectrum and didn't really think much about even why he was that way. But as time went on, I talked to my wife, and she said, 'No, no, no, he was like this and this'—there was a period when I wasn't there when he was just born, he was very alert, and if I know anything, I know her, and knowing our son—I know him so well myself—but she knows certain other things that I felt, I said 'You know, she might be right.'"

At the time of our interview, Bob and Grace's son was almost nineteen. "A wonderful kid, got a great sense of humor," says Bob.

I ask him what his response is to people who say the vaccine-autism link is settled science or "a disproven myth."

"I would say, *but then who settled it?*" De Niro replies calmly but with resolve. "How was it settled? Where is the science? as Bobby Kennedy says. Where is the science? Here's what *we* have from all these studies"—he gestures to the height of a large stack of studies printed out for the news conference—"And here's what *they* have," he says, gesturing a theoretical, smaller-size stack. "It seems like something is not right."

As part of the interview, I ask Bob, "Are you anti-vaccine?" I know the answer, but it's something the public should hear from his own lips after all the controversy. "No, I'm not anti-vaccine," he says. "That's like 'You're a witch' . . . the Salem Witch Trials. All of a sudden, you're 'anti-vax.' That's a lot of baloney. A lot of malarkey. It's ridiculous. I'm not anti-vax. I take vaccines all the time. And my kid's gotten vaccines. But there's something wrong. And it's gotta be fixed."

That pretty much ends the public record of the De Niros and vaccines. Bob's big idea for a world-class screening event never came to fruition. Neither did his idea for a documentary, congressional hearings, or whistleblower testimony from Dr. Thompson. Even Robert De Niro couldn't pierce the vaccine cartel.

After promising that he wouldn't be derailed on the matter, President Trump reportedly never addressed the possibility of a vaccine safety commission again after the initial conversation with Robert F. Kennedy Jr.

Bob and Grace announced divorce proceedings in 2018 after twenty years of marriage.

CHAPTER 8

For $ale

As an investigative reporter, understanding conflicts of interest at play in science and medicine is crucial. After confronting so many of them, I began to realize they don't only serve as important background for a news story—sometimes, they *are* the story. Such is the case in one of my reports for the *CBS Evening News* on July 25, 2008.

One year earlier, Rick Kaplan had been named executive producer of the *Evening News*. Kaplan was well aware of escalating pressure from both inside and outside of CBS to slant or halt reporting on prescription drug and vaccine safety. Some inside the network were growing wobbly under the strain. But Kaplan seemed to understand that no line of reporting was more impactful and important to our audience, and he wasn't one to be easily censored.

Meeting with me in my Washington, DC, office at 2000 M Street in early 2008, Kaplan tells me he wants me to propose a series of stories addressing drug and vaccine safety.

"I'll run them all," he tells me, indicating they would be shown on the *Evening News*.

"Are you sure you want to do that?" I ask.

"I'll run them all," he repeats slowly, for emphasis.

The first story in the series that we decide to do is a straightforward investigation into three of America's most ardent vaccine defenders:

the American Academy of Pediatrics, "Every Child By Two," and pediatrician Dr. Paul Offit. Each has strong but frequently undisclosed financial ties to the vaccine industry they promote. We title the news report, "How Independent Are Vaccine Defenders?"

The American Academy of Pediatrics (AAP) is the nation's largest professional association of pediatricians. Even today, its website makes the sweeping and misleading generalization that "[v]accines are both safe and effective." Surely this doctors' group knows that safety and effectiveness, as well as risk-benefit ratio, vary depending on the medicine and the patient. It's hard to imagine the AAP doesn't understand that vaccines pulled from the market due to safety concerns certainly aren't (and weren't) "safe and effective" for all. These withdrawn vaccines include the oral polio vaccine; Rotashield-brand rotavirus vaccine for diarrhea; plasma-derived hepatitis B vaccine; DTP (diphtheria, tetanus, pertussis) vaccine; and several flu shots. Also pulled for safety reasons: versions of vaccines for DTaP; measles, mumps, and rubella (MMR); haemophilus influenzae; Lyme disease; and rabies.

Though parents don't usually receive or read the packaging material, every vaccine approved in the US carries a list of possible adverse events that has ranged from brain damage, including autism, to death. There are also warnings describing which patients shouldn't be inoculated due to preexisting health conditions. For the American Academy of Pediatrics to encourage parents to blindly adopt all vaccines for all children rather than—at the very least—urge them to be on watch for any vaccine reactions is irresponsible and potentially very dangerous.

In my story for CBS News, I report that the American Academy of Pediatrics (AAP) accepts untold millions from the vaccine industry for conferences, grants, medical education classes, and even to help construct the group's Illinois headquarters. The doctors' group keeps the total amount secret, which seems contrary to the interest of transparency, but I manage to find some public documents that reveal bits and pieces. There's a $342,000 payment to AAP from Wyeth, maker of the pneumococcal vaccine, a jab worth $2 billion a year in sales at the time. There's a generous $433,000 contribution from Merck the same year AAP happened to endorse Merck's controversial HPV vaccine for cervical cancer, which had racked up $1.5 billion a year in sales. And another top donor: Sanofi Aventis, maker of seventeen vaccines and a new five-in-one combo shot that had just been added to the childhood vaccine schedule.

As for "Every Child By Two," at the time, it's a nonprofit that promotes early immunization for all children. It's often quoted by the media as if it's an independent child advocacy group. Yet, from what I see, its actions seem contrary to the health of the babies and toddlers: the group's stated priorities. *Why isn't Every Child By Two interested in learning about emerging vaccine safety concerns? Why aren't its advocates committed to fully investigating the questions and informing parents? Why did a spokesman from Every Child By Two once call CBS executives moments before I aired a vaccine safety investigation, in a frantic attempt to convince CBS to pull the story?* As with the American Academy of Pediatrics, when I ask Every Child By Two about its funding sources, the group won't tell me how much money comes from vaccine interests. A spokesman replies, "There are simply no conflicts to be unearthed." *Then, what's the big secret?* As I mentioned earlier, I looked at IRS tax forms filed upon the nonprofit's creation and in the years after. Listed as "Treasurers" for Every Child By Two are none other than officials from vaccine-maker Wyeth, as well as a paid advisor to big pharmaceutical clients.

Then there's Dr. Paul Offit, once perhaps the most widely quoted denier of vaccine harms. The CDC and vaccine industry allies often direct reporters to the supposedly independent Dr. Offit to be interviewed on vaccine safety questions. He can usually be counted on to reassure the public that any vaccine-related fears are unfounded. He's gone so far as to say that babies can tolerate "10,000 vaccines at once." When Dr. Offit gives interviews, he typically relates his title as "chief of infectious disease at Children's Hospital of Philadelphia and a professor of pediatrics at Penn's medical school." But that omits crucial context. As I report in my *CBS Evening News* investigation, Dr. Offit is a vaccine industry insider. He sits in a $1.5 million dollar chair funded by Merck at Children's Hospital. He holds the patent on RotaTeq, a controversial rotavirus antidiarrhea vaccine he developed with a Merck grant reported to be valued at $350,000. And around the time I air my story, future royalties for Dr. Offit's Merck vaccine have just been sold for $182 million. As part of my story research, I've asked Dr. Offit how much money he's made from RotaTeq. I've also asked for his conflict of interest disclosures—the list of pharmaceutical companies that have paid him. All publishing scientists should have that list easily at hand since ethics practices require them to disclose their financial ties when they publish. But Dr. Offit refuses to answer the questions.

Dr. Offit, along with the American Academy of Pediatrics, and Every Child By Two, all decline to be interviewed for my CBS report, but claim they're up front about the money they receive, and that it doesn't sway their opinions.

What happens after my story airs on CBS is even more telling about the dubious tactics of these players. Within a few days, Dr. Offit gives an interview to the *Orange County Register* to try to discredit my story and smear me. The newspaper article, titled "Dr. Paul Offit Responds," crosses my desk. In it, Dr. Offit claims that I "lied" in my CBS report. He says that I made my request for an interview with him in a "mean spirited and vituperative" email, and that I stated, "You're clearly hiding something." He also insists that he'd been completely cooperative and transparent, and had provided me with "the details of his relationship, and Children's Hospital of Philadelphia's relationship, with pharmaceutical company Merck."

Those were all invented fabrications on the part of Dr. Offit—more flagrant than I realized even he was capable of. At the time, I still had my emails showing I'd been exceedingly professional and polite in dealing with Dr. Offit, even after he'd flatly refused to provide the information that he told the *OC Register* he'd provided. In my emails to Dr. Offit, I'd respectfully appealed to him to reconsider, extended my original story deadline for him, and explained that the information would help my story be as fair and accurate as possible. I never wrote anything remotely resembling, *You're clearly hiding something.*

I provide these emails to the *Register* and ask the newspaper to issue a correction. They do. The correction publicly unmasks Dr. Offit's deception. It states, in part, "An *OC Register* article dated August 4, 2008, entitled 'Dr. Paul Offit Responds' contained several disparaging statements that Dr. Offit of Children's Hospital of Philadelphia made about CBS News Investigative Correspondent Sharyl Attkisson and her report. . . . Unsubstantiated statements include: Offit's claim that Attkisson 'lied'; and Offit's claim that CBS News sent a 'mean spirited and vituperative' email 'over the signature of Sharyl Attkisson' stating 'You're clearly hiding something.' . . . [D]ocuments provided by CBS News indicate Offit did not disclose his financial relationships with Merck. . . . [T]he network requested (but Offit did not disclose) the entire profile of his professional financial relationships with pharmaceutical companies. . . . The CBS News documentation indicates Offit also did not disclose his share of past and future royalties for the Merck vaccine he co-invented."

One way to tell that Dr. Offit holds a de facto position as propagandist at Children's Hospital of Philadelphia on behalf of the vaccine industry is that even after his dishonesty was exposed in the most public and humiliating way by the *Orange County Register* correction, he remained an employee in good standing at the hospital. The CDC and vaccine interests continued to promote his opinions on vaccine safety as if he were a credible source.

As far as I can tell, I was the only reporter asking Dr. Offit and other "experts" about their financial ties to the industry they so vigorously defended. I think my accurate reporting is one factor that eventually tarnished Dr. Offit's shine as a supposedly "independent" go-to for vaccine safety questions.

Therefore, when Covid vaccines entered the picture, pharmaceutical industry allies had to find another shill ready to step into Dr. Offit's duplicitous shoes.

Dr. Peter "Victim" Hotez

One would hope that Offit-level fabrications among public health officials are rare. After all, medical professionals have codes of ethics, licenses to maintain, and an oath to *primum non nocere,* or "first, do no harm." But the sad fact is that many doctors like Offit are willing to twist facts and attack those telling the truth.

On May 7, 2021, I'm targeted by another such doctor. I first learn of the attack when I see my name evoked in strange tweets on Twitter. Tennis icon Martina Navratilova tweets: "Sharyl clearly likes to stir things up in a criminal way.shame on her amd (sic) everyone else spewing this crap." *Criminal behavior? What on earth is she talking about?* I wonder. I begin to see similar comments from others. "@WaltDisneyCo and @ABC should fire [Sharyl] for instigating this," reads another. I don't work for ABC but regardless . . . *Instigating what?* From another: "I tweeted @SharylAttkisson and informed her that doxxing is a crime." "Doxxing" means to identify someone's personal information online in an intimidating way. I've never "doxxed" anybody.

I click around on Twitter and find what seems to be inciting the verbal assaults. It's a May 6 tweet by Dr. Peter Hotez at Baylor College of Medicine.

Like Dr. Offit, Dr. Hotez is often cited in the news media as if he's an independent expert on vaccine safety. Like Dr. Offit, Dr. Hotez is also a vaccine industry insider. As founding dean of the National School of Tropical Medicine at Baylor College of Medicine, Dr. Hotez is beholden to the same government-pharmaceutical industry purse strings and partnerships as are so many academic institutions. During Covid, Dr. Hotez worked with Merck on a team to develop a Covid vaccine and a "manufacturing platform" to get it to market on an accelerated timetable. He's known to frequently play the role of attack dog, going after scientists, journalists, parents, and advocates who research or speak to vaccine safety issues. And when exposed for putting out misinformation, or criticized for his attacks, Dr. Hotez frequently plays the victim.

Dr. Hotez's May 6 tweet about me makes ridiculous, concocted claims. At the time, he's on a media tour, claiming "anti-vaxxers" and white supremacists are threatening his safety. He falsely states that I've "endorsed an article" on my website that encourages people to attack him, and that I've compared him to "Mengele," an infamous

Nazi doctor during World War II who conducted unethical experiments on his victims. Nothing like that appeared anywhere on my website. I have no idea what he's talking about.

I tweet Dr. Hotez telling him to remove the false and defamatory accusations. Instead, he doubles down. I start to theorize that his vaccine industry masters might be the ones writing some of his tweets, and that he is truly ignorant.

At personal expense, I hire an attorney to contact Dr. Hotez. Our letter reads:

Dear Dr. Hotez:

This law firm is First Amendment counsel to Sharyl Attkisson. We have been retained to pursue redress for your May 6, 2021, false and defamatory accusations against Ms. Attkisson on Twitter. More specifically, you knowingly, maliciously, and falsely accused Ms. Attkisson to your approximate 161,700 followers of endorsing you being doxxed and comparing you to Mengele, thereby encouraging threats against you, in an article she authored. These accusations are false and defamatory per se, and damages to Ms. Attkisson's reputation will therefore be presumed as a matter of law. While you have taken down the offensive tweet—only after doubling and tripling down on your false claims—a takedown will provide you no cover under the law. To afford you an opportunity to mitigate the reputational harm you have caused to Ms. Attkisson, I hereby demand—as more specifically described below—that you post (and pin to your Twitter account) a complete retraction of your false accusations.

After the legal letter, Dr. Hotez publicly admits his mistake on Twitter and takes back his outlandish accusations. However, most of those who read the original false information will never see the correction.

Fast-forward to June 2023. Dr. Hotez is working overtime on what I consider to be a propaganda media tour. On this day, his job is to counter Robert F. Kennedy Jr.'s positions on vaccine safety, no matter how scientifically supported they might be. Kennedy has recently addressed the topic on the popular *The Joe Rogan Experience* podcast. Dr. Hotez calls Kennedy's positions "anti-vaccine disinformation." In response, Joe Rogan tweets, "Peter [Hotez], if you claim what RFKjr is saying is 'misinformation' I am offering you $100,000.00 to the charity of your choice if you're willing to debate him on my show with no time limit." Rogan's tweet is quickly viewed over 50 million times. Dr. Hotez sees it and demurs. He won't debate. That prompts Twitter owner Elon Musk to weigh in. "[Hotez is] afraid of a public debate, because he knows he's wrong," Musk writes. Other Twitter users start adding to the debate money pot. "I will add $150,000 to @joerogan's wager so now $250,000 can go to charity and the public can hear an open debate on an important topic," tweets financial industry billionaire Bill Ackman. "I'll add 500k," tweets celebrity Andrew Tate. Another $100,000 from liberal-billionaire-turned-vaccine-safety-activist Steve Kirsch. I'm not rich, but I add $10,000 to the kitty. Pretty soon, the offer tops a million dollars.

After Dr. Hotez's made-up story about me, there's good reason to wonder how many of the stories he tells are really as he describes. A case in point is his 2022 appearance at a Symposium on Infectious Disease in Children at a New York Sheraton hotel. (Just so you know, meetings, seminars, and Continuing Medical Education courses that are infectious disease–related are usually dominated by vaccine industry interests trying to sell physicians on their products and deflect from vaccine safety questions.)

During the symposium, Dr. Hotez takes to social media to imply that his life is in danger. "Scary #antivax protests in front of the NY Sheraton today for the Infec Dis in Children Symposium," he tweets. "Many thanks to Sheraton Hotel security for getting me out safe [*sic*] this morning." To hear Dr. Hotez tell it, he's risking his life to bravely speak at events promoting vaccines.

Two posts on Twitter call into question Dr. Hotez's version of events. "Are we living in an alternate universe?" tweets one person. "Peter was outside, smiling, posing for pictures with protesters." Attached to the tweet is photo evidence: Dr. Hotez, grinning and appearing very comfortable, just outside the Sheraton entrance, arm around a female protester, also smiling. The protester in the photo chimes in with her own tweet. "@PeterHotez disappointed to see your lies about our encounter," she writes. "When someone lies about the little things can they be trusted with the bigger things? @joshbucky thanks for capturing the moment of Dr. Hotez happy and Smiling."

Embarrassed by the Joe Rogan challenge in 2023, Dr. Hotez again goes on a media tour. As usual, the obedient press can be relied upon to help. "A prominent vaccine scientist says he was 'stalked' in front of home after Joe Rogan Twitter exchange," blares CNN's headline. Insider writes, "A vaccine scientist says being tag-teamed and dogpiled on by Elon Musk, Joe Rogan, and RFK Jr. all at once has been 'overwhelming.'" *Forbes* (under a "breaking news" title): "Vaccine Scientist Peter Hotez Says He Was 'Stalked' After Billionaires—And Joe Rogan—Urge Him To Debate RFK Jr." And Axios writes: "The incident—which ultimately resulted in individuals approaching the scientist outside his home—highlighted the potential risks for researchers and medical professionals using the platform, which saw a rise in hate speech after its acquisition by billionaire Elon Musk."

As I've advised with other news topics, when you see the same storylines repeated by multiple news organizations at the same time, often using the same wording and taking the same editorial position, you should suspect it may be part of an organized campaign. Organic, real news coverage isn't that homogenous. Did the complaints from a vaccine industry doctor whom most people have never heard of really deserve headline status at multiple national news organizations? Or was it the result of a well-executed propaganda plan?

The portrayal of Dr. Hotez as a serial victim has a distinct purpose. It's part of a strategy to characterize people who question vaccine safety as "violent" in order to get their speech curtailed or censored. Maybe they'll even be arrested and charged with crimes like some nonviolent January 6, 2021, attendees who found themselves raided at home, at gunpoint, by a small army of FBI agents in tactical gear.

Dr. Hotez refuses the million-dollar offer to defend his claims on Rogan's podcast and debate RFK Jr., even though he could have given the money to the charity of his choice. He could even have given the million to a nonprofit that's secretly doing the bidding of vaccine makers like Every Child By Two. Instead, he continues to draw attention to his claims that he's victimized by "anti-vaxxers." He whines that "anti-vaccine activists" harassed him at his Texas home. And he tells left-leaning MSNBC: "I offered to go on Joe Rogan but not to turn it into the Jerry Springer show with having RFK Jr. on."

"Vaccinate Your Family"

Much as Dr. Hotez assumed the role of Dr. Offit in the new Covid era in attacking those raising rational questions about vaccines, another entity I reported on in my 2008 CBS story, "How Independent Are Vaccine Defenders?" also underwent a Covid-era transformation. I'm referring to Every Child By Two, the nonprofit funded by the vaccine industry.

On December 8, 2020, three days before Pfizer's Covid vaccine was approved for emergency use in the US, Every Child By Two changed its name. It became "Vaccinate Your Family." *Why limit promoting vaccines for children when there's a whole universe of humans to vaccinate, and a growing number of vaccines to push?* According to Vaccinate Your Family's IRS filings, the group "maintains one of the largest social media presences of any pro-vaccine organization, seeing 11.9 million impressions in 2021, an increase of 218% over the previous year, and 209,000 engagements on posts and tweets."

Vaccinate Your Family is also incredibly well connected. The group says it's part of the World Health Organization's Vaccine Safety Net, "a network of sites and online destinations" with 2.1 million total users. In addition, Vaccinate Your Family gets taxpayer money from the CDC to promote vaccines to low-income women on assistance through the federal Women, Infants, and Children Nutrition program. Vaccinate Your Family also worked with food pantries and arts organizations in the Pilsen neighborhood of Chicago with "events, giveaways and clinics to promote vaccine confidence."

Some of the information promoted by Vaccinate Your Family can be classified as misinformation. For example, the group's website reads, "While a sore arm, or a fever and fussiness in your child after being vaccinated is never easy, it means that the vaccine is working by creating an immune response (protection) that may last a lifetime." This overgeneralization could cause parents to ignore signals of serious vaccine injury such as toxic myopathy, which can rarely lead to a severe muscle disorder called rhabdomyolysis, and even death. Additionally, many parents who say their children developed autism from vaccines commonly report that their babies and toddlers had extremely sore arms, fevers, and fussiness after their shots. Those symptoms aren't always to be disregarded as a sign that "the vaccine is working."

So who are today's shakers and movers behind Vaccinate Your Family? I looked up the group's IRS forms submitted in February 2023. They list Kirsten Thistle as treasurer. According to Thistle's biography, she's a media messaging specialist, formerly head of the division representing pharmaceutical companies at APCO Worldwide. APCO is a public affairs behemoth representing some of the biggest and most powerful companies in the global bio-pharma space. Thistle refers to herself as the "principal architect" of the baby vaccination campaign for Vaccinate Your Family's previous incarnation, Every Child By Two. She's had her finger firmly held to stop the bleeding on major vaccine safety controversies ever since. "Kirsten has significant experience in the vaccine space," reads her bio.

> Kirsten built and ran a coalition of vaccine manufacturers focused on addressing vaccine hesitancy and correcting misinformation about vaccines, including questions about links between vaccines (MMR and later thimerosal) and autism, SIDS, multiple sclerosis, and the safety of adjuvants, among others. Kirsten has also worked to raise awareness of teen immunizations, including those that prevent meningitis and HPV, and created an adult immunization communications platform for the National Foundation for Infectious Diseases. Kirsten has participated in and presented at the National Adult and Influenza Immunization Summit. She's

> worked in close collaboration with vaccine advocates and opinion leaders, including the **Children's Hospital of Philadelphia [home to Dr. Offit and Merck vaccine initiatives]** Vaccine Education Center, the American Academy of Family Physicians, the **American Academy of Pediatrics**, Voices for Vaccines, Sabin Vaccine Institute, PKIDS, the National Meningitis Association, the Autism Science Foundation and many others. [Emphasis added.]

Once you know that Kirsten calls the Autism Science Foundation one of the "vaccine advocates" that she closely collaborated with, you can toss out any notion that Autism Science Foundation is a true scientific nonprofit reporting fairly on vaccine-autism links. It's part of the club trying to dispel questions about vaccine safety.

Naturally, I found other big-name pharmaceutical interests operating in the background of Vaccinate Your Family. Among supporters listed on its website are the CDC; vaccine makers Pfizer, Glaxo, Novavax, Johnson & Johnson, Merck, and Sanofi Pasteur; Julie Gerberding (former director of the CDC, ex-head of Merck vaccines, and head of the Foundation for the National Institutes of Health); and the infamous Dr. Offit, who also happens to be a Vaccinate Your Family board member.

It's easy to see how anyone with opposing information hardly stands a chance of breaking through the morass. Put another way, these interconnections demonstrate how a core group of players has appendages that reach deeply into countless organizations where they exert broad influence over thought and policy on scientific topics.

Science Fiction

Those tentacles also reach deeply into fact-checking and media literacy organizations. Typically, their goal isn't really to check facts or help people find credible media sources. Instead, they seek to blanket the information landscape with a singular slant so that you'll be inundated everywhere you look with one uncontested view. These groups are cross-pollinated by a handful of common benefactors—such as

Google, Facebook, Craig Newmark of Craigslist, and other billionaire activists—and they validate and amplify each other's biased work.

High on that list is the nonprofit **First Draft**, founded at the start of the 2016 election cycle just in time to lead a new wave of narrative-weaving under the guise of fact-checking that was about to grip American media.

First Draft positioned itself as a resource for news media to turn to in the US and around the world. But like so many other groups, it serves to slant the information landscape on behalf of pharmaceutical and political interests. In my research, I traced the origin of the phrase "fake news" in its modern context to First Draft. (Yes, before Donald Trump successfully co-opted the term.) I'd thought it strange that it seemed like 99 percent of the time, First Draft's examples of "fake news" were conservative-leaning or pro-Trump, but never falsehoods and bias in the liberal press. I wondered why a legitimate fact-checking group would ignore egregious lapses when committed by the left. This behavior plucked the same cognitive dissonance I felt when doctors seemed to blindly defend a problematic medicine—and turned out to be working for the drugmaker. *What would I find if I followed the money with First Draft?*

First Draft's tax forms weren't yet filed online when I began looking for them in 2017, so I called the group directly and asked where their funding came from. A First Draft spokesman told me it was supplied by Google. Google was owned by Alphabet, which was run by left-wing billionaire Eric Schmidt, who was all in for Hillary Clinton for president. No coincidence that First Draft's take on "fake news" always seemed to benefit the Democrat position. Since then, First Draft has expanded to distributing misinformation on vaccines and other health issues, always taking the industry side.

Duke Reporters' Lab is also funded by Google and another left-leaning social media corporation: Facebook. Additional donors include the left-leaning Knight Foundation, which provides grants for journalism, and (again) left-wing activist Craig Newmark of Craigslist. Duke Reporters' Lab works with other left-leaning fact-checkers, including

The Washington Post, PolitiFact, the International Fact-Checking Network, and FactCheck.org.

A leviathan in this space is the **Poynter Institute**. Poynter describes itself as a "nonprofit media institute and newsroom." It was a small but reputable news school when it was founded in 1975. However, the direction of Poynter changed radically around 2015, about the same time Google started up First Draft. And that's no coincidence. There were coordinated, global efforts to pour billions into the effort to create and expand left-leaning and pro-pharma fact checks and other initiatives to influence the 2016 election and stories the media would report.

Like Duke Reporters' Lab, Poynter is also funded by Google, Facebook, and Craig Newmark. In 2015, Poynter got $1.38 million from Newmark and the vaccine-promoting Bill & Melinda Gates Foundation "to influence news coverage of global health initiatives." In 2017, Poynter got $1.3 million from the left-leaning social change groups Omidyar Network and the George Soros–founded Open Society Foundations. In 2018, Google gave Poynter another $3 million to launch "MediaWise," an effort "to teach 1 million American teenagers how to sort fact from fiction online" and to reach out to first-time voters in the 2020 election. In 2019, Newmark announced another $5 million gift to Poynter. In 2020, Poynter used funding from Facebook to expand MediaWise into a voter project "to reach 2 million American first-time voter college students, helping them to be better prepared and informed for the 2020 elections."

Sadly, Poynter has allowed itself to be turned into a pro-pharma, left-leaning political and social activist operation under the guise of a journalism group. Its fingerprints can be found everywhere! Today, a lot of misinformation on health issues can be traced to Poynter and its supporters.

Columbia Journalism Review used to conduct important watchdog reporting on the pharmaceutical industry and conflicts of interest. One example was the article "Bitter Pill," written by Trudy Lieberman. But today, it often reads like a typical left-leaning misinformation rag, and Craig Newmark funds this publication, too. In February 2019, the Columbia Journalism School accepted a $10 million gift from the Craig

Newmark Philanthropies. Newmark also sits on the board of *Columbia Journalism Review.* Steve Coll, dean of Columbia Journalism School and Henry R. Luce Professor of Journalism, said, "Craig Newmark's generosity will provide an enduring and deeply influential investment in journalism. . . . At a time of disinformation campaigns and attacks on journalists online and off, the Center and faculty chair will send a powerful message and will bolster a free and ethical press that enhances our democratic society." Newmark issued a statement reading, "Both Columbia Journalism School and Poynter are already helping journalists do just that, and with these gifts, I hope they'll become the industry's go-to resources for the challenges journalists face in a data-driven world." It's remarkable how those intent upon shaping information, censoring off-narrative views and facts, and in some cases promoting disinformation, speak as though they're doing the opposite—and nobody challenges them.

Health Policy Watch touts itself as "a nonprofit, open-access journal" that provides "Independent Global Health Reporting." But that's not what it does. It's an activist website for left-leaning takes on issues such as climate change, and for pro-pharma positions on vaccines. *Health Policy Watch* partners with "journalists" to spread propaganda about "top global health policy issues." It operates with generous support from the left-leaning Wellcome Trust, which frequently partners with the familiar, vaccine-promoting Bill & Melissa Gates Foundation. The editor in chief of *Health Policy Watch* is Elaine Ruth Fletcher, a left-leaning climate change activist who wrote left-leaning articles for the World Health Organization (WHO). The group's number two, Kerry Cullinan, came from the left-leaning "Open Democracy," another activist group that promotes left-leaning gender and climate-change journalism.

Health Feedback and its umbrella group Science Feedback are prolific purveyors of disinformation. In my view, it looks like they were created for this very purpose. Like many similar groups, they distribute propaganda—often incomplete or untrue—about vaccines and other scientific matters under the guise of fact-checking and policing misinformation. They have been embraced and heavily relied upon by Facebook, Instagram, and TikTok, to name a few. Funding has come from Google, Facebook, pre-Musk Twitter, climate activist groups, and other left-wing advocacy organizations. From what I can tell, they always adopt positions on controversies that line up with these funders and the pharmaceutical industry.

My factually accurate reporting on various vaccine matters has drawn frequent attention from Health Feedback. In April of 2024, Health Feedback apparently found a five-year-old investigation I conducted (that "went viral") so compelling and potentially damaging to pharmaceutical interests, its staff dredged it back up and made it the subject of one of their full-blown fact checks. Of course, the fact check couldn't point to anything inaccurate in my story. It recounted the sworn affidavit signed by noted pediatric neurologist Dr. Andrew Zimmerman. As I described earlier, Dr. Zimmerman had been serving as one of the government and vaccine industry's chief expert witnesses in dispelling vaccine-autism links when he says he came to learn otherwise: that vaccines can cause autism after all, in certain children. He says the Department of Justice lawyers he told fired him as an expert witness and went on to misrepresent his professional opinion in court. Health Feedback set up a litany of straw man claims that I'd never made, and falsely claimed the whole story was debunked and discredited.

In March 2024, an uncharacteristic admission comes from a doctor operating an ironically named fact-checking website she calls "MisinformationKills." Dr. Allison Neitzel has spent recent years disparaging independent physicians who were off the government-pharma narrative on Covid. Among others, she targeted Drs. Paul Marik and Pierre Kory, founders of Front Line Covid Critical Care (FLCCC) alliance. Her criticisms included a study by Dr. Marik on the effect of vitamin C on sepsis, and a meta-analysis by both doctors on the use of ivermectin to treat Covid. But now, Dr. Neitzel posts a lengthy, public apology that sounds like it's being made to avoid legal action. It reads in part: "I used terms like fraud and fraudulent to criticize certain studies by Dr. Marik or Dr. Kory. My posts have also characterized the use of ivermectin in treatment with words like 'grift.'" She continues, "I regret if anyone understood the statements as accusations that any of them had engaged in fraudulent professional or business practices." She goes on to note why her critiques of certain studies by Drs. Marik and Kory were flawed or unfounded, and concludes by stating, "I apologize to the FLCCC, Dr. Marik, and Dr. Kory for the statements that are the subject of this update." After the apology, another smeared physician chimes in on X (formerly Twitter). Dr. Jay Bhattacharya of Stanford writes that Dr. Neitzel "defamed me, @MartinKulldorff, and @SunetraGupta by falsely insinuating that we received Koch Foundation support for the @gbdeclaration signed by thousands of scientists advocating against Covid shutdowns. This is categorically untrue, yet she has not retracted or apologized for the smear."

There are thousands of similar groups. If you look through their websites and mission statements, they typically claim they're independent and nonpartisan. They insist their donors have no influence over their fact checks and reports. But keep in mind they have no obligation to tell the truth. Their work product reveals their true colors. They use money from vested interests to write propaganda in the name of science. They coordinate with news organizations to distribute the propaganda and call it journalism. They are not unbiased scientific authorities no matter how many times they claim to be or how many reporters treat them as if they are.

CHAPTER 9

Along Came Covid

When President Joe Biden got Covid twice in July 2022, his second and third known Covid infections, he became the most public living symbol of the lapses, confusion, and government misinformation that mark America's long Covid nightmare.

Biden—along with top advisor Dr. Anthony Fauci; the head of Pfizer, Dr. Albert Bourla; and CDC Director Dr. Rochelle Walensky—all had insisted the vaccines would prevent Covid. By the end of October 2022, they counted at least nineteen shots among them, and nine bouts with Covid. Dr. Walensky, five vaccinations into her personal Covid prevention program, had just battled two cases of Covid in one nine-day period, with the CDC refusing to answer my questions about exactly when she'd had her shots, or whether she'd had Covid before. *Why the secrets?*

On March 29, 2021, in an appearance on MSNBC, Dr. Walensky famously and falsely proclaimed, "Vaccinated people do not carry the virus. Don't get sick." I was dumbfounded. Much of the world already knew from personal experience that the CDC director was dead wrong on this point. Vaccinated people *were* getting ill with Covid in large numbers, and would soon outpace the number of unvaccinated people sick from Covid. How could it be that nearly everyone knew better than the head of the world's most prestigious infectious disease authority?

No matter. With science misrepresented at the highest levels, a conflicted news media was all too happy to perpetuate the false information rather than question it or even tell both sides. The left-leaning magazine *Fortune*, purveyor of slanted and false information on numerous topics, blasted out an untrue, unqualified headline: "It's official: Vaccinated people don't transmit COVID-19." According to the media, whatever a favored public figure utters is "official" and true, even when it's patently false.

Even within a sea of government disinformation, this newest Walensky flub stands out as really, really bad. Not to mention easily disproven on its face. That must be why Dr. Walensky's own agency quickly issued a sheepish-sounding correction. A CDC spokesman understated, "It's possible that some people who are fully vaccinated could get COVID-19." Years later, when I checked, the false *Fortune* headline remained. "Vaccinated people don't transmit COVID-19."

The more the government, media, and other vaccine industry interests tried to discredit Covid counter-voices, the more they discredited their own positions. In fact, propaganda that promoted Covid-vaccines-at-any-cost arguably did more to undermine confidence in America's vaccination program and public health authorities than anything I've witnessed. Examined this way, it becomes clear how destructive the wholesale attempts to control "science" and medical information has become.

Much has been written about the Covid postmortem. A lot of it has been inaccurate or missed the point. This chapter examines ten key moments among a voluminous number of misrepresentations made to the public by once-trusted authorities. Together, these "moments" illustrate how deceptive modern science has become, and explain far better than anything else why we find ourselves in such critical condition.

1. The Truth About Isolating at Home

There was strong, early evidence that public health recommendations to "isolate at home" actually spread Covid! We now know with certainty that infected people, some of them without symptoms, stayed indoors, sharing breathing space with family for long periods of time, *increasing* the odds that they would spread the virus to loved ones. Independent scientists like Dr. Jay Bhattacharya of Stanford, who advocated against Covid shutdowns from the start, rightly said that people should have been encouraged to spend time outside, visiting parks and beaches. By the time our federal health authorities admitted scientists like him were correct, the erroneous guidance had taken a large toll.

An early hint that it was a mistake to instruct people to lockdown and stay in close contact with each other at home comes on May 6, 2020. New York governor Andrew Cuomo makes a little-examined comment that catches my ear. At the time, New York is suffering the first big Covid spike in the US. Cuomo delivers what he calls shocking news: the vast majority of people sick in the hospital with Covid had isolated themselves just as the government asked! To be specific, Cuomo says early data from one hundred New York hospitals reveals that two-thirds, or 66 percent, of new Covid-related admissions are people who had largely sheltered at home—most of them retired or unemployed.

Shouldn't this revelation have made international news and guided us to quickly reconsider stay-at-home orders?

Cuomo goes on to add that another 21 percent of the hospitalized patients came from group living facilities, mostly nursing homes that were practicing strict Covid controls, including isolation and masking. So that makes a total of 87 percent among the seriously ill who had *not* gone out to grocery stores, shopping malls, parks, beaches, restaurants, or the office. They were not riding in taxis and Ubers, or on the subway. They were not at concerts and parties, or going out to visit friends and family.

"This is a surprise," Cuomo remarks. "Overwhelmingly, the people were at home. We thought maybe they were taking public transportation, and we've taken special precautions on public transportation, but, actually no, because these people were literally at home."

The remarkable data seems to fall on deaf ears. It's not what the government and medical establishment want to hear. They've already loaded the American public onto a train barreling irreversibly toward a chosen destination. So they continue to urge people to stay home. And when it comes to the venues that are actually the safest, such as outdoor parks and beaches: they're shut down. How many people got ill or died because public health officials dismissed hard data that was evident from the start?

An exclamation to this point comes with the famous annual Sturgis Motorcycle Rally. In August 2020, 460,000 bikers descended upon the Black Hills of South Dakota at the height of the coronavirus pandemic, making the rally one of the year's most-criticized events in the media. Another reason it fell under such criticism is that a lot of the attendees were supporters of President Trump. The media and politicians quickly labeled it a "superspreading" event. One "scientific study" claimed the Sturgis rally was responsible for more than a quarter million Covid-19 cases, an astonishing 19 percent, or nearly one in five, of all cases reported in America at the time! To me, the wild claims weren't credible on their face. I'd learned to trust my cognitive dissonance and decided to investigate further. Why weren't hundreds of journalists questioning the obviously questionable narratives instead of blindly furthering them?

In early 2021, I set off for Sturgis. It's a charming small town of about 7,000 people in western South Dakota. It's flanked on the south by the craggy mountains of the Black Hills National Forest. It's a cold, cloudy day and the Sturgis I see when I arrive is pretty quiet. Not at all like the town looks when the throngs come for the rally. I connect with city official Daniel Ainslie for an interview. He begins by recounting outlandish anecdotes of people who'd been counted as "Sturgis" Covid cases back in their own home states even though they'd never stopped anywhere near Sturgis or the rally!

"We had one individual that stated that they were just driving to Washington State, and they were driving along I-90, which of course runs through our community, and so then they were counted as one of the Sturgis recipients, even though they didn't even stop in Sturgis," says Ainslie. "According to their state health official, apparently they were a Sturgis victim of the coronavirus."

Ainslie tells me that in advance of the rally, a contingent of assorted special interests warned town officials that if they didn't cancel the 2020 rally, models predicted Sturgis hospitals would be overwhelmed, and up to 5 percent of people in town would *die*!

Fortunately, nothing even close to that happened.

The truth is, there's no way science could tell if *anybody* was infected at Sturgis, let alone who and how many. But a lot of hard data shows how wrong the alarmist media reporting and "scientific" projections proved to be.

First, based on statistics that health officials publicized and used at the time, any group of 460,000 people should have seen several hundred Covid deaths among them. However, even the hyperbolic media reports ultimately linked Sturgis to only between one and no more than five possible fatalities. That's an incredibly low number—and those were pure guesswork. Also, none of those deaths was scientifically traced to the rally; they were just the only ones the media could find in anyone who had supposedly been to the event.

Second, community-wide mass testing in Sturgis after the rally found a mere 26 Covid-positive cases out of 650 people tested. Again, that's a far, far lower rate than most anywhere else.

Third, news reports about Sturgis that were intended to sound upsetting actually proved to be reassuring, when critically examined. For example, a *New York Times* article stated: "In all, [Sturgis Motorcycle Rally] cases spread to more than 20 states and at least 300 people—including revelers' families and co-workers who never set foot in South Dakota." But again—only 300 cases among 460,000 attendees? That works out to an infection rate of just .065 percent, or about six-hundredths of 1 percent. That would make the Sturgis rally one of the safest places in America—not a superspreading event!

In summary, to try to claim that there were a quarter million Covid infections from Sturgis, as a San Diego State University IZA study had done, was "ridiculous . . . fanciful, and it was just pushing their narrative," observes Ainslie.

Far too late, the CDC tacitly acknowledged that its recommendation to isolate indoors was a mistake. The agency eventually noted on its website, "Spending time outside when possible instead of inside can also help [prevent Covid spread]: Viral particles spread between people more readily indoors than outdoors." Mayo Clinic wrote more plainly: "It's much harder to catch the virus . . . when you are outside." And the EPA now states, "Being indoors rather than outdoors" increases the risk of infection.

2. Fauci's Follies

Did Dr. Anthony Fauci of the National Institutes of Health (NIH) have some diabolical intent behind his many misstatements and falsehoods? Was he beholden to unseen masters due to conflicts of interest? Or was he simply utterly befuddled? Whatever the case, he became one of the most important disinformationists of the pandemic, yet was never held accountable.

The die was cast against Dr. Fauci's credibility as early as March 2020, in the first weeks of America's pandemic, when I heard him tell one of his biggest whoppers under oath.

As director of the NIH's National Institute of Allergy and Infectious Diseases, and the lead Covid advisor to the White House, Dr. Fauci repeatedly cited jarring and scary figures in public. On March 11, 2020, while testifying to Congress, he stated that Covid-19 was "ten times more lethal than the seasonal flu." The alarming testimony generated international headlines that blared across the Internet and television news, and remains frequently cited today.

I was skeptical. I'd read far more conservative assessments from some of Dr. Fauci's learned colleagues in academia and even, as it turns out, from Fauci himself!

"[T]he case fatality rate may be considerably less than 1%," Dr. Fauci wrote in an article two weeks later, published in the *New England Journal of Medicine* (*NEJM*) on March 26, 2020. "This suggests that the overall clinical consequences of COVID-19 may ultimately be more akin to those of a severe seasonal influenza (which has a case fatality rate of approximately 0.1%) or a pandemic influenza (similar to those in 1957 and 1968) rather than a disease similar to SARS or MERS, which have had case fatality rates of 9 to 10% and 36%, respectively."

Why was Dr. Fauci contradicting his own public testimony and giving a far softer take on Covid's potential to kill, when writing in a scientific journal? Here he was saying Covid's death rate was *like* a bad flu season, not ten times deadlier as he'd just claimed to Congress.

To confuse matters further, just one day after the *NEJM* article, I found Dr. Fauci back to repeating the higher fatality number rather than "considerably less than 1%."

"The mortality of [COVID-19] is about 10 times [flu]," Dr. Fauci tells *Comedy Central* host Trevor Noah on March 27, 2020.

Though Dr. Fauci was never shy about granting television interviews, from local news and the networks, to CNN and Fox, he wouldn't agree to an interview with me for my Sunday television news program *Full Measure.* So I posed the simple questions to his office via email: *Which of his accounts was accurate? Was Covid's death rate similar to that of the flu, or ten times deadlier?*

Dr. Fauci didn't answer. And among members of Congress or journalists who were able to get close to him, I never heard the questions asked. Nor was the bizarre discrepancy in Dr. Fauci's accounts widely scrutinized in the press.

There are many other examples of Dr. Fauci making conflicting claims. One that received some media attention was his flip-flop on masking to prevent the spread of Covid. After originally saying masks wouldn't work, he became part of the effort to impose strict mask mandates and shame those who chose to show their faces. He even advocated wearing more than one mask at a time! Yet, you may recall he was caught on video at an outdoor sporting event, sitting maskless close to his friends. One could speculate that Dr. Fauci knew all along that masks wouldn't stop Covid from spreading, despite what he told the public.

In an email later obtained through the Freedom of Information Act, Dr. Fauci had said as much to a friend who'd apparently asked if she should wear a mask. It was early in the pandemic, on February 5, 2020.

"The typical mask you buy at a drug store is not really effective in keeping out virus, which is small enough to pass through the material," Dr. Fauci told his friend. "I do not recommend that you wear a mask."

In fact, Dr. Fauci seemed to get it wrong on most every critical point involving Covid-19. He shamelessly promoted the experimental and unproven vaccines without reservation and, as I mentioned earlier, joined vaccine makers in furthering the false narrative that Covid vaccines would prevent infection. For example, on December 12, 2020, he claimed the vaccines had "been found to be up to 95 percent effective in preventing the COVID-19 illness."

A year later, he was still banging the same drum in the face of all evidence to the contrary. "We do have interventions, in the form of a vaccine to prevent infection," Dr. Fauci insisted, even though it's well known the vaccines do not prevent infection.

3. Mortal Mistakes

Calculating how deadly a disease is—its fatality rate—is mathematically simple. It requires fourth-grade skills. Take the number of deaths (numerator) and divide it by the total number of people who caught the disease (denominator). The denominator should include people who got sick as well as those who were infected but didn't actually have symptoms, or "asymptomatic" cases. So let's say 100 people get OGREVIRUS, and two of them die. That's 2/100. The death rate for OGREVIRUS is 2 percent. Simple.

But I noticed early and often during the Covid pandemic that officials routinely made bush-league calculation mistakes. And somehow, the mistakes always resulted in exaggerating Covid's dangers.

Here's one way that happened. According to the CDC at the time, most people infected with Covid had few or no symptoms and were unlikely to get tested. But as I mentioned, their numbers still needed to be represented in the total disease group, or denominator, in order to calculate an accurate death rate. Omitting the people who had Covid, perhaps asymptomatically, but didn't get tested would make the denominator too small and result in the death rate seeming far worse than it really was. Yet officials routinely excluded these people! Instead, they used only the subset of people who'd tested positive for Covid. This error had the impact of greatly overstating Covid's fatality rate.

By way of example, let's return to OGREVIRUS. Assume there are 100 people, and 40 get sick (testing positive), and 60 are infected without symptoms (and never take a test). Next, assume 2 die. That makes the fatality rate 2/100, or 2 percent. But what if you made the mistake of calculating the fatality rate using 2 dead among only the 40 sick people who went and got tested instead of all 100? The OGREVIRUS death rate jumps from 2 percent to 2/40, or 5 percent! Now OGREVIRUS appears to be two and a half times deadlier than it really is!

That's exactly what public health experts did over and over again with Covid fatality calculations—unchallenged by their colleagues, analysts, and the media. I found it mystifying that I seemed to be the only reporter to note this fallacy. I began searching scientific literature to confirm whether my rudimentary math exercise was, in fact, accurate. I discovered an article on this very topic cowritten by Drs. Eran Bendavid and Jay Bhattacharya of Stanford University. They rightly pointed out that "the true fatality rate" should be calculated among *all* infected people, not just the subset of those identified through a positive test. They commented that, as I'd observed, the common miscalculation done with Covid was resulting in startling but inaccurate figures. For example, Italy's much-publicized Covid fatality rate of 8 percent was estimated using confirmed cases, rather than total infections. Drs. Bhattacharya and Bendavid pointed out, Italy's "real fatality rate could, in fact, be closer to 0.06%."

4. More Math Tricks

The more people who get vaccinated, the less Covid there should be—*if the vaccines work*. Don't let anyone double-talk you into thinking there's something more complicated to it. Yet as millions quickly got vaccinated in 2021, the trend line went in the wrong direction. Covid infections rose to record levels, far outpacing the pre-shot numbers. You don't have to be a scientist to understand what a disappointment the Great Vaccine Hope turned out to be. In fact, it quickly began to seem possible that vaccinated people were *more likely* to get Covid.

On August 6, 2021, Dr. Walensky and the CDC were once again

bringing up the rear. They finally acknowledged, publicly, that fully vaccinated people who catch Covid can infect others, after all—something they'd long denied. There's no legitimate explanation for why the media often refused to report headline-worthy studies reflecting this fact, such as one published by Harvard and Canadian scientists in September 2021. It found that the more highly vaccinated a population was, the more Covid cases there were. According to the scientific analysis:

> [C]ountries with higher percentage of population fully vaccinated have higher COVID-19 cases per 1 million people. Notably, Israel with over 60% of their population fully vaccinated had the highest COVID-19 cases per 1 million people in the last 7 days. The lack of a meaningful association between percentage population fully vaccinated and new COVID-19 cases is further exemplified . . . by comparison of Iceland and Portugal. Both countries have over 75% of their population fully vaccinated and have more COVID-19 cases per 1 million people than countries such as Vietnam and South Africa that have around 10% of their population fully vaccinated.

The numbers here in the US were equally concerning. In fall of 2021, about three in ten Covid deaths (30 percent) were vaccinated people, according to the Kaiser Family Foundation. By January 2022, that rose to about four in ten (40 percent). Then, by April 2022, CDC data showed that most deaths, six in ten (60 percent), were among vaccinated people.

Every time hard data undercut the government and vaccine industry messaging, a convincing new narrative had to be invented. It would require some deft sleight of hand to explain away this trend and keep people on board with an endless parade of equally ineffective boosters. So the final master narrative about vaccines was put on the table. *See, the reason more vaccinated people are getting sick and dying of Covid is that there are* more *vaccinated people,* the propagandists told us. *It's basic math! You're not stupid, are you?*

"More vaccinated people are dying from COVID, but that's due in large part to a larger percentage of the whole population being vaccinated & the effectiveness of boosters waning," read a typical tweet on the topic. Another tweeter who tried to explain away the high number of deaths among the vaccinated wrote, "There are many more vaccinated people."

But for that explanation to make sense, the explainers would have to be indirectly admitting vaccines don't work in the first place. If the shots are *not* effective, then yes, as more people are vaccinated, there will be more deaths among vaccinated people. But if the vaccines actually *work*, the larger vaccinated crowd would make up a far smaller proportion of Covid deaths.

I saw this ridiculous "more-people-are-vaccinated" rationale deployed in a coordinated fashion by propagandist publications such as *Scientific American*, and amplified by many seemingly intelligent people. To them, the nonsensical seemed to make perfect sense.

Here's another numbers trick that a lot of people seemed to fall for. It was the CDC's practice of giving the impression that it was very, very dangerous to be unvaccinated. For example, CDC's website boasted a frightening claim: unvaccinated people were nearly 13 times more likely to die from Covid! The agency also routinely publicized other studies that sounded terrifying. One covered "25 jurisdictions" and found "the death rate was 53 times higher among the unvaccinated than the boosted." "People have a much higher chance of dying from COVID if they're not vaxxed," warned a typical tweet in response.

But even if we accept the CDC's chosen interpretations and its most-amplified numbers at face value, it still glosses over an important but unstated takeaway: *Covid's overall death rate was proving to be minuscule, regardless of whether the person was vaccinated or not.* Why wasn't *that* a headline—that Covid rarely kills? And why wasn't Covid's low lethality balanced against the risk of relatively untested vaccines with serious side effects, short and long term, known and as yet unknown, when given over and over? Why was the CDC always seeming to choose the most alarming public health message that promoted the experimental vaccines, instead of providing balanced, rational, and measured advice?

5. Natural Immunity Deniers

One of the worst failings by the CDC and public health officials was their abject denial of the power of natural immunity. Here, too, propagandists manipulated statistics for their own purposes.

Unlike the CDC, many independent scientists correctly predicted early on that natural immunity after Covid would work better than any vaccine. This is how it's always been with illnesses and vaccines. Even Dr. Fauci initially touted the power and hope of natural immunity before deciding to almost singularly promote an endless string of Covid vaccines and boosters.

When natural immunity's superiority over Covid vaccines became evident in the scientific literature, vaccine industry interests worked to engineer data to make it seem as if vaccines were more effective than they were.

By way of example, assume this scenario that ignores natural immunity: Out of 100 people, half are vaccinated and half are unvaccinated. Forty-five of the 100 get Covid in the newest wave. Among the 45: 15 were vaccinated but 30 were not. In other words, 15/50 of the vaccinated got Covid. But 30/50 of the unvaccinated got Covid. The vaccine looks highly effective! It's twice as dangerous to be unvaccinated!

But that's only if you pretend natural immunity from infection doesn't exist.

Out of the same 100 people, consider this possibility: What if all 55 of the people who didn't get Covid in the newest had natural immunity from a prior infection? And what if none of the 45 people who got Covid had a prior infection? Suddenly, it's natural immunity that emerges as the powerful defining factor, and the vaccine becomes irrelevant. By not considering natural immunity, the CDC made it impossible for ordinary folks to factor in one of the most important protective factors of the pandemic.

Based on what we know, we can assume that if natural immunity hadn't proven to be so effective, the CDC would have publicized numbers demonstrating that. The fact that the agency pretended the question was unimportant gives away the truth.

Late in the game, the CDC quietly acknowledged that people who'd had Covid were, indeed, protected by natural immunity. But the agency still used phrasing that would tend to promote vaccinations, anyway. On its website in November 2022, the CDC posted: "If you become ill with COVID-19 *after you received all COVID-19 vaccine doses recommended for you,* you are . . . considered up to date [emphasis added]. You do not need to be revaccinated or receive an additional booster."

6. Word Games

Everybody knows what it means to be vaccinated. At least we used to. The failure of Covid vaccines to prevent infection, transmission, symptoms, or serious illness gave rise to some serious wordsmithing by vaccine industry interests. They sought to redefine current meanings and invent new ones. They *had* to, in order to keep the masses in a state of suspended disbelief, complying with increasingly questionable vaccine requirements.

Once upon a time, there were two simple-to-understand terms: "vaccinated" and "unvaccinated." However, the Pfizer and Moderna RNA vaccines required two shots in the initial series. A person might have gotten one shot but, we were told, that didn't count as "fully vaccinated." (As an aside, this means the CDC found a way to skew their statistics and dismiss the many people who got Covid after their first shot but before the second. They weren't counted as vaccine failures because they weren't "fully vaccinated" at the time.)

Next, a whole lot of people were still getting Covid shortly after the second shot. So now, we were told, *those* people weren't "fully vaccinated" until two weeks passed after the second shot. This adjustment, again, magically prevented untold numbers of people who got Covid within two weeks of their two-shot series from getting counted in the vaccine failure tally. You see, *they weren't fully vaccinated.* In fact, it masked the very real possibility that vaccination was making people more likely to immediately catch Covid.

Then, as more "fully vaccinated" people continued to get Covid more than two weeks after their second shot, the goalposts were moved again. We were told that to be considered fully vaccinated, you actually needed a booster in addition to the two shots and the two-week wait. As more people with boosters got Covid anyway, they needed yet another booster to be considered fully vaccinated. And another.

Meantime, most everyone got Covid at least once, vaccinated or not. The illness rarely became serious or fatal, and the vaccines became virtually meaningless in terms of impact. But rather than give up, vaccine industry interests doubled down. They decided we should get annual Covid vaccines. So the CDC announced its intention to shelve the term "fully vaccinated" altogether and replace it with "up to date." That means your obligation to continue getting vaccinated in order to supposedly be protected—never ends.

It's not just the concept of "fully vaccinated" that needed a revamp in order to shoehorn reality into the pro-pharma agenda. When the Covid vaccines failed to meet the definition of "vaccines" because they clearly didn't prevent the disease, there was a successful campaign to literally redefine the word "vaccine." Why admit a product is a failure when you can unilaterally change the meaning of a word and claim it's a success? So in early September 2021, on CDC's web page, somebody made the change. "Vaccine" had been defined as "A product that stimulates a person's immune system to produce immunity to a specific disease, protecting the person from that disease." But the phrase "protecting the person from that disease" was removed. Today the CDC says that vaccines merely "stimulate the body's immune response." This Orwellian redefinition of two hundred years of the world's understanding of what constitutes a vaccine was executed without so much as an explanation, public discussion, or vote.

7. Elderly Risks Ignored

Early in the nation's Covid vaccination program, I made it my mission to do some original research. It wasn't easy to find time, with my full-time television news program and a serious obligation to teach tae kwon do and train for my fifth-degree master's black belt. But I knew that if I was going to be positioned to speak authoritatively to important, emerging trends, I'd have to dig in and do some deep dives. A lot of that was accomplished in the quiet hours in the middle of the night. I'd sit at my computer with background noise on the television losing track of time, trying to defeat Google-forced search results in order to find information that powerful interests don't want us to see.

As I conducted my research, it seemed to me that many health officials and journalists were recirculating the same conjecture, quoting the same dubious sources, and not doing their own independent fact-finding to confirm or cast doubt on what we were all being told. The worst part was how many medical professionals, reporters, politicians, and "fact-checkers" were claiming to definitively know things that they couldn't possibly know. Meantime, independent researchers and other voices were being bullied and censored if they asked logical questions or presented facts that challenged medical establishment narratives.

I'd already learned that one of the easiest ways to find raw, unadulterated data is to comb through the online federal database called VAERS, the Vaccine Adverse Event Reporting System. It is something I've done countless times over the years when looking for side effect trends.

Looking through the VAERS entries at the start of the Covid vaccine push, I quickly identified two small clusters of deaths in elderly people after they were vaccinated at nursing homes in Kentucky and Arkansas. These cases should have raised eyebrows at the CDC and garnered great attention in the scientific community. But they didn't.

In Kentucky, four seniors died the same day they got the Pfizer-BioNTech vaccine on December 30, 2020, according to VAERS reports. Three of the four reportedly already had Covid prior to getting vaccinated. In other words, the vaccine was of no known benefit and could even prove harmful, but they were given it anyway. Victim number one was an ill eighty-eight-year-old woman described as "14+ days post covid." Someone had given her the shot while she was "unresponsive in [her] room." She died within an hour and a half. The second death was that of a man, ninety, who was "15 days post covid." He was given the shot and died within ninety minutes. A third report says an eighty-eight-year-old woman who was "14 + days post covid" got her shot, vomited four minutes later, became short of breath, and died that night. And an eighty-five-year-old woman vaccinated at 5 p.m. was "found unresponsive" less than two hours later, and died shortly after.

In response to my questions about the Kentucky cluster, a spokesman for the CDC said its experts noted "no pattern . . . among the [Kentucky] cases that would indicate a concern for the safety of the COVID-19 vaccine."

In Arkansas, four more seniors died at a long-term care facility about a week after vaccination. All four had gotten the Moderna vaccine. All tested positive for Covid after vaccination! There's no indication as to whether they had Covid when they were vaccinated or acquired it after their shots. Three of the four patients got their shots on December 22, 2020, and died about a week later. The person reporting the case of an eighty-two-year-old man who died six days after his shot said he'd been vaccinated in an attempt to "mitigate his risk" and that "this was unsuccessful and [the] patient died." Two elderly women, ages ninety and seventy-eight, got vaccinated on January 2, 2021, tested positive for Covid about a week after their shots, and died. The unnamed person who reported the ninety-year-old's death speculated, "the vaccine did not have enough time to prevent COVID 19," and added the ridiculous and unsupported comment, "There is no evidence that the vaccination caused patient's death. It simply didn't have time to save her life." A fourth patient at the same facility was a sixty-five-year-old man who was also vaccinated on January 2. He died two days later.

In response to questions about the Arkansas cluster, the CDC told me, "Surveillance data to date do not indicate excess deaths among elderly patients receiving COVID-19 vaccinations."

What makes these deaths even more significant is something that Norway was discovering very early in its vaccination program. There, 23 people also died shortly after vaccination. After investigating 13 of the deaths, Norway's medical agency concluded that common side effects from the Pfizer and Moderna vaccines, such as fever, nausea, and diarrhea, "may have contributed to fatal outcomes in some of the frail patients."

"There is a possibility that these common adverse reactions, that are not dangerous in fitter, younger patients and are not unusual with vaccines, may aggravate underlying disease in the elderly," said Steinar Madsen, medical director of the Norwegian Medicines Agency. Of course, this should have come as no surprise. We've already discussed how vaccination or other medicine can exacerbate a preexisting vulnerability. In Norway, they paid attention to the possible danger signs and recommended that the frail and elderly avoid the shots. In the US, our health officials explained away the warning signs.

In combing through VAERS reports, I found numerous other early cases of elderly, frail people in the US who'd had Covid, got vaccinated, and died. A ninety-six-year-old Ohio woman tested positive for Covid in November 2020, got the Pfizer vaccine anyway in a rehab facility after a fall a month later, and died that afternoon. A ninety-four-year-old Michigan man at a senior living facility who had Covid and other illnesses got the Moderna shot anyway on January 2, 2021, and died of cardiac arrest two days later. A ninety-one-year-old Michigan woman with Alzheimer's and other illnesses at a senior living facility who'd tested positive for Covid got the Moderna jab anyway on December 30, 2020. She died four days later. And an eighty-five-year-old California woman with Alzheimer's and other disorders at a senior living facility had asymptomatic Covid but was given the Pfizer BioNTech vaccine anyway on January 5, 2021. She was found dead the same day. A 104-year-old woman in New York got the Pfizer vaccine on December 30, 2020. The next day, a Covid test was done and came back positive. She became ill the following day and died on January 4, 2021. A seventy-one-year-old New York man was given the Moderna vaccine on December 21, 2020, developed a fever and respiratory distress, and tested positive for Covid. He was given the government-endorsed injectable medicine remdesivir. He died after six days.

A World Health Organization vaccine safety subcommittee reviewed reports of deaths among the frail and elderly after the Pfizer vaccine and assured the public that there was no cause for concern.

I wonder how many lives might have been lengthened if we'd taken greater care to examine the risks of vaccinating the frail and elderly, or if we'd understood the consequences of vaccinating older people who'd already had Covid. Put differently, how many lives were cut short?

8. The CDC's Big Lie

What makes the case of elderly deaths after vaccination most concerning is that the victims were all vaccinated shortly after the CDC disseminated important but shockingly false information that may have cost them their lives. The CDC wrongly claimed the original vaccine studies definitively showed the shots somehow benefited people who'd already had Covid. *But the studies didn't show any advantage to people who'd already had Covid.* It wasn't even a main research question addressed during the rushed studies. So there was no good reason to vaccinate the nursing home patients who'd already had Covid.

Further, it could be argued that some of the scant data gathered early on indicated that, if anything, previously infected people who got vaccinated were at greater risk of getting Covid. For example, in Pfizer's study, for test subjects who'd already had Covid, the vaccine had an effectiveness of minus 17.9 (–17.9 percent). In simpler terms, they were *more likely* to get Covid again after vaccination according to some scientists who reviewed the study. And in Moderna's original study, test subjects who'd had Covid, then got vaccinated anyway, were six times more likely to have to drop out of the study because they suffered such serious side effects.

On its website, the CDC publicized its disinformation about Pfizer and Moderna studies supposedly finding that vaccination provided benefits to people who'd already had Covid when none had been proven. The agency distributed the false information in its *Morbidity and Mortality Weekly Report.* The inaccurate claims were signed by the CDC's esteemed group of vaccine advisors. Remarkably, none of them seemed to know or care that the CDC's misrepresentation of the original vaccine studies was untrue and could prove harmful.

It wasn't Pfizer, Moderna, the government's top health experts, public health officials, medical reporters, or "fact-checkers" who unearthed the CDC's error. They all just parroted and passed around the same bad information. *Had any of them actually read the studies?* It was a member of Congress who exposed the truth. Representative Thomas Massie, a Republican from Kentucky, identified the CDC's disinformation in December 2020, while researching whether he should get vaccinated after he'd already had Covid.

As part of his personal query, Massie ended up secretly recording an astonishing series of phone calls with a string of duplicitous CDC officials and scientists. When pressed by Massie on the calls, they seemed to reluctantly admit that they knew the vaccine studies didn't show what the CDC had publicly claimed. Yet their conversations with Massie were filled with double-talk, and the officials balked at the idea of publicly correcting their misinformation. In January 2021, I published these revelations and the audio recordings in a story for *Full Measure*.

In the phone calls, Massie can be heard pressuring the CDC to tell the public the truth: that the original vaccine studies did not show any benefit to people getting the shots if they'd already had Covid. Shockingly, just days after one recorded conversation in which CDC officials finally agreed to make corrections to the public record, the same CDC officials took part in an online webinar for medical professionals *where they repeated the misinformation.* In that presentation, the CDC's Dr. Sarah Oliver reiterates the false claim: "Data from both [vaccine] clinical trials suggests that people with prior infection are still likely to benefit from vaccination."

For weeks, in a series of phone calls, Massie kept up the pressure on the CDC to tell the truth. The CDC eventually issued a halfhearted correction. I say halfhearted because it used language that would still lead most people to falsely believe the original studies showed the vaccines benefit people previously infected with Covid. *What's their motivation? Why was CDC so stubborn about telling the truth?* One possible answer: admitting the vaccines were unnecessary for people who'd had Covid would mean a vastly shrunken market for vaccine makers. After all, Covid was racing through the population at breakneck speed. If all those people wouldn't benefit from vaccination or, worse, could be harmed by the shots, the vaccine bonanza would be over in an instant.

Another crucial question that all of this raises is why did the CDC's top advisors—its committee of so-called vaccine expert advisors—all sign off on the false information in the first place? Are they corrupt or just ignorant? The CDC never explained. Nor did it hold anyone accountable.

Meantime, two very large, outcome-based published studies from Cleveland Clinic and Israel in 2021 expanded the knowledge base. They concluded there's no scientific reason to vaccinate people who already have natural immunity from Covid. Vaccination exposes them to risks with no benefit.

This story is meaningful on several levels. If top CDC scientists and officials are intentionally distributing false information, what does that imply about the advice they're giving and policies they're making concerning the many vaccines we give our children, or other disease challenges we face, and the trillions of tax dollars we've provided to address these problems?

9. All the Fake Claims

Rather than a single key moment, this category comprises a group of common Covid-related misrepresentations. Many propagandists have attempted to revise history and claim these things were never said. These specific examples prove otherwise.

On November 9, 2020, media reports claimed one of the Covid vaccines "was found to be more than 90% effective in preventing COVID-19 in participants."

On November 18, 2020, CNN reported that "Pfizer's coronavirus vaccine is 95% effective in preventing Covid-19 infections, even in older adults, and caused no serious safety concerns."

In December 2020, vaccine makers crowed about their products' supposed ability to prevent infection: "Moderna's initial Phase 3 clinical data in December 2020 was similar to Pfizer-BioNTech's—both vaccines showed about 95% efficacy for prevention of COVID."

On December 12, 2020, Dr. Fauci claimed the vaccines have "been found to be up to 95 percent effective in preventing the COVID-19 illness."

On February 12, 2021, an article in *USA Today* claimed Covid vaccines "were all 100% effective in the vaccine trials in stopping hospitalizations and death."

On March 16, 2021, Pfizer announced that "its COVID-19 vaccine is extremely effective at preventing both symptomatic and asymptomatic cases of the disease."

On March 29, 2021, CDC claimed the vaccines were 90 percent effective at preventing infection.

On March 31, 2021, Becker's Hospital Review reported that "Pfizer says its vaccine is 100% effective in preventing COVID-19 in adolescents."

On April 1, 2021, Pfizer chairman and CEO Albert Bourla tweeted, "Excited to share that updated analysis from our Phase 3 study with BioNTech also showed that our COVID-19 vaccine was 100% effective in preventing #COVID19 cases in South Africa. 100%!"

Also on April 1, 2021:

- A study in the *British Medical Journal* claimed the vaccines prevent infection, as well as symptoms.
- *Fortune, New York* magazine, *The Intelligencer*, MSNBC, and many others claimed Pfizer and Moderna shots are highly effective at preventing infection and illness.
- CDC Director Dr. Walensky claimed, "Vaccinated people do not carry the virus—they don't get sick."

On June 7, 2021, a CDC study claimed "the mRNA COVID-19 vaccines . . . reduce the risk of infection by 91 percent for fully vaccinated people."

On June 14, 2021, an article in *Science* claimed that Novavax, Pfizer, and Moderna vaccines were "essentially 100% protective against disease."

On July 6, 2021, a Stanford study claimed vaccines prevented Covid infection in nearly 100 percent of people.

On August 27, 2021, the CDC claimed, "The Pfizer-BioNTech and Moderna mRNA COVID-19 vaccines were approximately 90% effective in preventing symptomatic and asymptomatic infection" of Delta variant.

On September 23, 2021, *The Atlantic* claimed vaccinated people are less likely to spread Covid.

On October 3, 2021, Dr. Fauci stated, "we do have interventions, in the form of a vaccine to prevent infection."

On October 7, 2021, President Biden claimed people who are vaccinated "cannot spread it."

On October 11, 2021, a study by a group of Indian scientists claimed four out of five vaccinated people won't get Covid.

On November 22, 2021, Pfizer issued a press release claiming its vaccine demonstrated "100% efficacy against COVID-19 in longer-term analysis, with no serious safety concerns identified" in children ages twelve to fifteen.

On December 14, 2021, President Biden claimed that vaccinated people "do not spread the disease to anyone else" and "This is a pandemic of the unvaccinated."

In January 2022, justices on the Supreme Court displayed shocking ignorance of facts and played a remarkable role in furthering disinformation—even basing decisions on it. At the time, the court was hearing opposition to the Biden administration's sweeping and unprecedented Covid vaccine mandates. Here is some of the incorrect information repeated by the justices.

From Justice Sonia Sotomayor:

- Omicron was as deadly or deadlier than Delta
- 100,000 children were in critical care, many on ventilators, due to Covid
- Covid deaths were at an all-time high at the time

From Justice Elena Kagan:

- It's "beyond settled" science that vaccines stop Covid transmission

- It's proven that masks are highly effective at stopping the spread

From Justice Stephen Breyer:

- Vaccine mandates would prevent 100 percent of Covid cases in the United States
- 750 million people tested positive for Covid-19 on one day that week. (The US population is roughly 330 million people.)

10. Miscounting Covid

One of the biggest and most calculated misrepresentations of the pandemic has to do with how the public health establishment conspired to miscount Covid deaths. There was a time when a news organization with massive resources like the *New York Times* would have deployed reporters across the country to examine death records and find the true facts. Yet there was no major effort to scratch beyond the surface.

I was able to get at the heart of the matter by visiting an American county where Covid deaths were so few and far between, the coroner was in a unique position to flag any miscounts. It was Grand County, Colorado, about a hundred miles outside of Denver.

It's a chilly, sunny day when I knock at the door of the inauspicious offices of Grand County Coroner Brenda Bock, who has agreed to an on-camera interview with me. She doesn't look like a coroner, though I'm not certain what a coroner should look like. She comes off as someone's friendly, smart, great-aunt. She takes me on a brief, winding tour of the small offices, and then we begin our conversation.

Bock starts by telling me that in November 2020, she got the sad news that on Thanksgiving Day, a troubled young man named Lucais Reilly had shot his wife, Kristin, in the head, then committed suicide. Both were in their thirties and had alcohol and drugs in their system, along with a history of domestic troubles. Then something strange happened.

"The very next day [the deaths] showed up on the state website as Covid deaths," Bock tells me. "[But] they were gunshot wounds. And I questioned that immediately because I had not even signed off the death certificates yet, and the state was already reporting them as Covid deaths."

Bock says some unknown authority had apparently learned about the deaths, run the couple's names through a database, found they'd tested positive for Covid sometime within twenty-eight days prior to their deaths, then recorded them as Covid deaths even though they were a murder-suicide. Miscounting this poor couple as Covid deaths had the impact of helping to tilt national Covid stats to give the scary misimpression that Covid was killing young, healthy people.

Within a week of the murder-suicide, Bock says, two more Grand County deaths popped up on the state's Covid count. But she says she had no record of the deaths. So she looked into it.

"[They] were actually still alive," she says. "And yet [the state was] counting them." When she called state officials to challenge the deaths, she says she was told, "Oh, well, that was a typo. They just got put in there by accident."

The mysterious miscounting was so concerning that Bock and Grand County's commissioners joined to write a letter to Colorado's governor asking him to correct it. The trend wasn't just happening in Grand County, according to the chief medical examiner and coroner for Denver, Dr. James Caruso. I also visited with Dr. Caruso on my trip to Colorado. He did an interview with me in his Denver morgue, which was steely and sterile. He was proud to show it off.

"I was told by some of my fellow coroners in the more rural counties in Colorado that it was happening to them," Dr. Caruso tells me, referring to non-Covid deaths being counted as Covid. "They knew of issues where they had signed out a death certificate with perhaps trauma involved, and they were being advised that it was being counted as a Covid-related death. At some level—maybe the state level, maybe the federal level—there's a possibility that they were cross-referencing Covid tests and that people who tested positive for Covid were listed as Covid-related deaths, regardless of their true cause of death. And I believe that's very erroneous, and not the way the statistics needed to be accumulated."

I asked Colorado governor Jared Polis for an interview about the

Covid death miscounts for my TV report. He didn't agree to talk with me, but a spokesman told me the governor agreed with Coroner Bock's complaints and was "outraged" that a murder-suicide is recorded as Covid-related. Yet Bock says the governor personally told her he didn't intend to remove the cases from the Covid count because "all the other states were counting deaths the same way."

Amid the controversy, Colorado officials did make one adjustment to their methodology. And it ended up shedding some light on the magnitude of the miscounts. They split Covid deaths into two categories: those who really died of Covid, and those who died of something else but happened to test positive for Covid within a few weeks of their death, like the murder-suicide. What impact did this have? In summer of 2021, the state's total Covid-related death tally was 13,845. But subtracting the deaths not directly caused by Covid cut the fatalities in half!

As I continued my research, I also found traffic fatalities counted as Covid deaths, an alcohol poisoning counted as a Covid death, and three Colorado nursing home deaths attributed to Covid, even though the attending physicians specifically said the cases weren't related to coronavirus. One egregious miscount covered on the local news in Nashville, Tennessee, was the case of Hal Short. He tested negative for coronavirus three times right before he died of an aggressive cancer. Yet Covid-19 was listed as the cause on his death certificate. It was only changed after the family repeatedly complained. "How many other people are you making this mistake with?" Short's widow asks in a news report about the scandal.

I found official death counts that the media relied upon were just as flawed. When I checked in July 2021, the *New York Times* Covid death tally overreported Grand County's 2020 Covid death toll by at least 500 percent. The *Times* was missing one resident who reportedly died of Covid outside of the county. But the *Times* mistakenly counted an unrelated heart attack as a Covid death, as well as the two people who Bock found were still alive (who had since been removed from the official state total), and the murder-suicide of Lucais and Kristin Reilly.

Colorado coroners I spoke to are convinced the state was bent on overcounting Covid deaths and they provided examples of that. But the system for collecting the information was such a mess, one could just as easily argue the theory that the state undercounted Covid deaths. As it stands, nobody can really know for certain. But it sure seemed to many observers that there was an effort to hype the numbers.

Not many localities seemed to be concerned with getting an accurate death count. They were perfectly happy to inflate the numbers with non-Covid deaths, and the feds were all too eager to accept them. One exception was Alameda County, California. In June 2021, Alameda County officials announced a change in methodology to remove deaths from its Covid count that weren't a direct result of Covid. That reduced their death toll by 400 people, or 25 percent. The federal government never did take steps to filter out non-Covid deaths from the total count at the national level. Yet that step would be crucial to understanding the true cost of the pandemic, and who was most likely to get the sickest.

With these examples in mind, it becomes easy to see how virtually every aspect of "science" during the Covid pandemic was manipulated. The public was asked to swallow a constant flood of disinformation pablum that was forced down our throats by experts we were supposed to trust the most. These experts betrayed their public obligations, violated ethical tenets, sullied their profession, and stole taxpayer salaries and grant money while serving industry masters.

But few breaks in public trust were more consequential than the case of hydroxychloroquine.

CHAPTER 10

Following the Money with Covid Treatments

If you consume the curated, mainstream medicine and media narrative, you're probably under the impression that the prescription drug that President Trump touted as a possible game changer against coronavirus—hydroxychloroquine—has been thoroughly "debunked" and discredited.

At the height of the Covid panic—as politics, money, and medicine intersected—two divergent views of hydroxychloroquine emerged, both of which could not possibly be true at the same time. The first view was the negative one widely reported in the press. The second view you've likely heard less about. Rarely has a discussion about choices of medicine been so polluted by political overtones.

Before the coronavirus vaccines hit the market, there was a desperate search for an existing medicine that might help. Studies from China and France sparked early hope that hydroxychloroquine might work against Covid-19. The drug had been on the market for decades, long proven safe and effective at preventing mosquito-borne malaria. Pre-vaccine, President Trump rightly escalated the need to study hydroxychloroquine's potential against coronavirus.

From a traditional scientific viewpoint, it made sense. When a novel disease breaks out, it's long been established that the first and fastest line of defense is to see if existing, approved medicine might be effective. But with Trump's first endorsement of hydroxychloroquine, there was a major media-driven effort to portray it as dangerous quackery. Suddenly, a drug that had been used safely and without controversy since the 1940s was portrayed as risky and even deadly! The media campaign against hydroxychloroquine was assisted by a highly publicized online report published in mid-April 2020. It claimed that for sick Covid patients treated under the US Department of Veterans Affairs, hydroxychloroquine did not help and, even worse, was linked to increased deaths.

An independent news media would have done more than just parrot the spin on an emerging question to which nobody could rightly claim to know firm answers. An independent news media would have dug into the research and looked at various views, including those of the scientists who theorized hydroxychloroquine could be an effective tool. Instead, most reporters eagerly adopted the designated narrative. The anti-hydroxychloroquine story pushed out for public consumption was designed for two purposes. First, it would build demand for the soon-to-be-marketed multibillion-dollar vaccines. Second, it would further discredit Trump at a time when that was already Job One for many in the media. President Trump was running for reelection from a surprisingly strong position. That was even after being framed by a false story fabricated by Democrats and the FBI about "Russia collusion," and after being impeached by Democrats over an equally questionable narrative about his actions and policies related to Ukraine. The media's assignment now was to dispel the notion that a cheap, available drug might help prevent or treat Covid. If an existing drug could do that, it would eliminate the justification to rush out experimental and unproven vaccines on an emergency basis.

On April 6, 2020, the *New York Times* was among those that falsely reported "Trump . . . seized on [hydroxychloroquine] as a miracle cure." Trump never said that.

"Why are you promoting this drug?" a reporter demands at one of Trump's Coronavirus Task Force briefings, referring to hydroxychloroquine.

"I'm not," replies Trump.

"You come out here every day, right, sir? Talking about the benefits of hydroxychloroquine?" retorts the reporter.

"I want them to try it," says Trump. "Now, it may not work, in which case, 'Hey, it didn't work' . . . And it may work, in which case it's going to save a lot of lives."

At the time, hundreds of Americans were said to be dying of Covid. The CDC and the medical community at large, funded by trillions of tax dollars over the decades to be prepared for just such a scenario, seemed at an utter loss when it came to a quick, effective plan of action. Trump was pulling out all the stops to jump-start solutions.

"This [Veterans Affairs] study showed that actually more people died that used the drug than didn't," argues the reporter who's challenging Trump. His statement disregards other credible research that showed a benefit to using hydroxychloroquine.

While most of the press is busy dumping on hydroxychloroquine, I detect a simultaneous, concerted effort to rally popular support around a newer medicine: remdesivir. Remdesivir was first developed for Ebola, but never approved by the FDA. Early tests of remdesivir in coronavirus patients supposedly showed it helped them recover four days faster, but with no "survival benefits." In other words, it didn't save any lives. Another problem with remdesivir is that it's delivered as an IV fluid in the veins, generally limiting its use to late in the game for very sick patients in a hospital setting who may be too far gone to help. Compare that to hydroxychloroquine, a cheap pill that some studies indicated could prevent serious illness if taken early on—or perhaps even prevent infection in the first place!

But the beleaguered hydroxychloroquine gets demonized, while White House advisor Dr. Fauci seems almost giddy about remdesivir. "That the data shows that remdesivir has a clear-cut significant positive effect in diminishing the time to recovery—this is really quite important," Dr. Fauci excitedly tells the TV cameras. From then on, camps largely divided along political lines . . . and it had nothing to do with scientific evidence. Many right-leaning media figures and independent scientists sided with hydroxychloroquine, while the public health establishment and left-leaning press backed remdesivir. Each accusing the other of ignoring real science.

I found an informative view from cardiologist Dr. William O'Neill, a medical director at the Henry Ford Health System in Detroit, Michigan, where researchers were studying both remdesivir and hydroxychloroquine and had no known financial stake in the outcome.

"I've never seen science so politicized in forty years of practice," Dr. O'Neill tells me when I visit with him.

"Some people in the media are treating hydroxychloroquine as if it's something that's being pitched by charlatans [as if] it's dangerous, debunked, and discredited," I say to Dr. O'Neill. "What do you make of that?"

The question and scenario remind me of a conversation I'd had in May 2008 with Dr. Bernadine Healy, former head of the National Institutes of Health. At the time, I was interviewing her about the supposedly debunked link between vaccines and autism—which hadn't been debunked, at all. It was an issue many scientists were coming to learn was nuanced, though presented to the public as black and white.

"I think you can't say that [it's been debunked]," Dr. Healy tells me about vaccines causing autism. She goes on to say that public health officials have intentionally avoided researching the critical question of whether some children are "susceptible" to vaccine side effects, afraid the answer will frighten the public.

"There is a completely expressed concern that they don't want to pursue a hypothesis because that hypothesis could be damaging to the public health community at large by scaring people," Dr. Healy explains. "I don't think you should ever turn your back on any scientific hypothesis because you're afraid of what it might show."

It's shocking and brave that she's willing to expose the dirty little secret she says some of her colleagues are conspiring to keep.

Now, more than a decade later, Dr. O'Neill makes an observation similar to Dr. Healy's when he talks about demonization of hydroxychloroquine.

"I think [the negative media narrative on hydroxychloroquine] is very harmful," Dr. O'Neill tells me. "President Trump touted it early, and so then the media set out to disprove and discredit it without any regard for science. I think those of us that are actually involved in the scientific endeavor feel that there is some value to it, and it has to be tested."

Dr. O'Neill is a world-renowned leader in interventional cardiology, famous for "pioneering research in new techniques to diagnose and treat heart attacks." He's not only an eminent researcher; he's also a clinician. When we first spoke, he'd already prescribed hydroxychloroquine, before the government limited its use, to numerous Covid patients. He tells me *he saw improvement in all of them.* Real-world, firsthand evidence directly from a scientist involved! On the other hand, he tells me he's far less impressed by remdesivir. "There's a lot of hype for [remdesivir]," Dr. O'Neill continues. "I saw the original *New England Journal* article study and I saw the *Lancet* study, and to me it's just like a big 'Ho-hum.' I just don't see a big benefit."

Adding to the public drama and confusion is a draft version of a study that apparently wasn't supposed to be published but accidentally did get posted on the Internet. It showed that remdesivir didn't help most Covid patients and caused serious side effects. The results were so troubling that the eighteen test subjects involved in the study were taken off the drug. The maker of remdesivir, Gilead, didn't respond to my query about that, but has said publicly that it ended the controversial study because it couldn't find enough volunteers to take part.

Against this backdrop, on May 1, 2020, the FDA announces a big decision that furthers the prevailing narrative against hydroxychloroquine and gives an edge to the experimental remdesivir. The FDA allows emergency use of remdesivir for severely ill coronavirus patients. At the same time, it restricts hydroxychloroquine. With the change, the FDA says that hydroxychloroquine can only be used in the hospital or as part of a formal study due to reports of "serious heart rhythm problems."

The FDA's clever maneuver has the impact of eliminating hydroxychloroquine's potential advantage as early help for people before they get too sick. That's because the FDA has put hydroxychloroquine on equal footing with the weaker but government-favored remdesivir, which, as an IV drug, is generally limited to hospital use. Now hydroxychloroquine can only be used in the hospital too.

So the questionable remdesivir gets hearty endorsements from establishment medicine, while new restrictions on hydroxychloroquine mean it will almost never be used. Studies were making it clear that hydroxychloroquine may work well if given early, but not on hospitalized, seriously ill patients. With the FDA's new maneuver, in the rare cases where it will be given in the hospital, it will be too late. That will skew any reviews of its worth, making it look like it just isn't effective.

Independent scientists begin asking: *Who benefits from the government taking away a medicine that could prevent serious illness, and guaranteeing more people will be admitted to hospitals for expensive treatments that might fail?* The answer is vaccine makers and their partners in government. That's because under unambiguous wording in section 564 of the Federal Food, Drug, and Cosmetic Act, experimental vaccines *can only be used in an emergency when no existing drugs work*: "For the FDA to issue an Emergency Use Authority, there must be no adequate, approved, and available alternative." Like hydroxychloroquine or ivermectin.

As the hydroxychloroquine drama is playing out in real-time, an intriguing contact reaches out to me. He's a scientist and medical doctor who's quietly advising the Trump White House. He's been following my reporting on the hydroxychloroquine controversy and tells me that *I'm onto something.* He wants to meet in person. This is at the height of the Covid shutdowns and most people aren't venturing out into the world. But based on what I've learned through my reporting, I'm okay with risking exposure to real life. I'm heeding the advice of well-placed scientists who indicate that, from what we know so far, going about normal activities is the best Covid strategy for healthy people to follow. The White House advisor I'm meeting up with apparently feels the same. We agree to rendezvous in the parking lot of J. Gilbert's wood-fired grill in McLean, Virginia, between my Leesburg, Virginia, home and Washington, DC.

The day of our meeting, the restaurant is shut down, of course. The world is already steeped in the global economic disaster that would grip us for many years to come. So when I pull into the restaurant parking lot, I rightly assume the only other car there belongs to my source. It's an expensive little sports car, and the man in the driver's seat motions for me to get into the passenger side. I climb in and shut the door, and we greet one another in close quarters with the windows rolled up. It's not the sort of thing most people would do at the time. He tells me not to worry.

"You're safe," he assures me. "I'm on hydroxychloroquine."

He proceeds to tell me his opinions about the public tug-of-war over the drug. He's been urging the Trump White House to go all-in on investigating hydroxychloroquine as a treatment for and possible prevention of Covid. But he says he's finding himself up against a strong contingency in the Fauci–Big Pharma camp.

"Yes, hydroxychloroquine can cause heart problems," he tells me, addressing the much-publicized safety concerns. "But only at levels far beyond what any human would ever take." He tells me that *he* should know—because he's using himself as a guinea pig. A researcher and medical doctor, he'd taken doses of hydroxychloroquine much higher than recommended for Covid, until he stimulated a heart effect in himself, and backed off.

"If you take quadruple the proper dose, it could be problematic," he tells me. "But nobody is suggesting people take anything close to that for Covid."

After the sports car encounter, I speak to a third scientist (after Dr. O'Neill and the White House advisor) who also makes the argument that hydroxychloroquine is being unfairly disparaged. She's Dr. Jane Orient, head of the Association of American Physicians and Surgeons (AAPS). AAPS doesn't accept pharmaceutical industry money and bills itself as an independent alternative to the Big Pharma–supported American Medical Association (AMA). Dr. Orient's group has fallen under vicious attack due to its independence.

"How do you account for the difference in medical and scientific opinion about [hydroxychloroquine]?" I ask her. "Because some people seem so certain that it can be a positive benefit to coronavirus patients—maybe even crucial in the early days—whereas some people are convinced it should absolutely not be used."

"That's a very good question," Dr. Orient replies. "But the ones who have the most experience are very enthusiastic about the possibilities. And we do have naysayers that we suspect may have a little conflict of interest because they are so enthusiastic about remdesivir, which is a new drug that hasn't been approved for anything, and that, so far, is showing really very equivocal or even negative results." Largely escaping the media's attention was research like a July 15, 2021, study that concluded "routine use of remdesivir may be associated with increased use of hospital beds but not with improvements in survival."

Dr. Orient hints at financial interests behind the narratives. "I think we have to look at the money," she tells me in an interview for *Full Measure*. "There's no big profits made in hydroxychloroquine. It's very cheap, easy to manufacture, been around for seventy years. It's generic. Remdesivir is a new drug that could be very expensive and very lucrative. So I think we really do have to consider there's some financial interest involved here."

A recurring lesson in science, and life in general, is that when the actions of media, government, and public officials seems so contrary to common sense: *Follow the Money*. You can learn a lot. So, amid the hydroxychloroquine nonsense, I start connecting the dots. First, I look to see who was behind that quickly produced, highly publicized Veterans Affairs report in April 2020 that was so critical of hydroxychloroquine. I find out that one of the authors received research funding from Gilead, the maker of remdesivir. The funding included a major $247,000 grant in 2018.

Next, I start wondering about the government panel of experts empowered to devise the coronavirus treatment guidelines. I know from my past reporting that medical advisory panels are often rife with conflicts of interest. Panel members who land those influential positions often work for, or have been paid by, companies that benefit from whatever the panel decides. I reported on a prime example of such conflicts of interest in 2004.

Cholesterol Conflicts

The cholesterol controversy begins back in 2004 when a panel of eight experts issues startling, new government guidelines claiming our cholesterol levels need to be much lower than we thought. And the panel of experts—all doctors—say that more of us need to begin taking shiny, new, cholesterol-lowering statins. At the time, statin drug sales have already hit $19 billion a year. The new guidelines, under the National Institutes of Health National Cholesterol Education Program (NCEP), create eight million more potential customers for statins overnight!

But to many independent observers, the advice seems radically unbalanced. It pushes patients toward medicine that has serious side effects ahead of safer, impactful diet and exercise changes. Some suspect deception. They begin looking into the background of those eight experts who advised the National Cholesterol Education Program. None of them had disclosed any financial ties or conflicts of interest related to their work. But it turns out they were hiding something big. Seven of the eight doctors were linked to statin makers that benefited from the new guidelines!

Dr. John Abramson of Harvard Medical School is one of the medical sleuths who made this disturbing conflict-of-interest discovery "with just a little bit of research," as he tells me. Upon the discovery, he and three dozen independent scientists sign a petition urging the National Institutes of Health to reject the questionable new cholesterol guidelines and seek an "objective, independent" review. They suggest such a review would likely "lead to different conclusions" than the ones reached by the doctors paid by statin companies. After the conflicts of interest are exposed and questioned, the National Institutes of Health belatedly admits that seven of the advisers have financial ties to twenty-nine drug companies, but rejects the petition to toss out their advice.

"Of course [the conflicts of interest] should have been disclosed" before the cholesterol guidelines were released, Dr. Abramson tells me in an interview. "But the much more important thing is those people shouldn't be writing guidelines." Yet the government allows and facilitates it. This incentivizes drug companies to buy up nearly every expert who's likely to be consulted on matters regarding a medicine.

Hydroxychloroquine: Same Old Story

Which brings us back to hydroxychloroquine. Sixteen years after the statin lesson, I go down the list of names on the government's Covid treatment panel, which dialed back hydroxychloroquine use and promoted newer, more costly remdesivir. I'm looking to see if any of them are connected to companies that make hydroxychloroquine or remdesivir. Ideally, none of the panel members should have links to companies benefiting from their decisions. But that world no longer exists in public health. A second-best option would be for advisors to recuse themselves from decisions involving companies they're connected to. But that world doesn't exist either.

What do I find? About one-third of the government's Covid treatment advisory panel, or eleven members, reported having links to a drug company. Nine of the eleven named Gilead, the maker of remdesivir! *What are the natural odds that so many people chosen for an expert panel on Covid would be related to the company making the treatment they ended up favoring?* I keep digging. Beyond the eleven, I identify nine others, including two of the committee's three leaders, who also have ties to Gilead but hadn't disclosed them. Two even served on Gilead's advisory board! Others were paid consultants or received research support and honoraria from Gilead. They weren't technically required to fess up those relationships because, under the committee's special rules, for some reason, panel members don't have to acknowledge conflicts of interest older than eleven months. *As if conflicts of interest somehow become irrelevant after eleven months?*

There's one more piece of the puzzle I need to find. How many on the same advisory panel report having financial links to remdesivir's disfavored rival, hydroxychloroquine—a generic drug so cheap that no big drug company will make billions if it works on Covid?

The answer: zero.

Hydroxychloroquine's Demise

When I first spoke to Henry Ford Hospital's Dr. O'Neill, he remained confident that good science would prevail, and studies underway would soon provide definitive answers on hydroxychloroquine's effectiveness for Covid.

"I think that it's just still very early in this disease process that we're going to learn a lot," he tells me in late spring of 2020. "There's six hundred studies that are being done in the United States right now on Covid to see all sorts of different kinds of infections and combinations. We're going to be a lot smarter at the end of the summer. So I think what I would just say to everybody, just hold your powder."

Not long after, I learn that Dr. O'Neill's hydroxychloroquine study has been halted by unnamed authorities above his pay grade. It had become pointless, anyway. The FDA's criticism of hydroxychloroquine had made it impossible to get enough volunteers for the study to move forward. After all, who in their right mind would raise their hand to take part in a trial with medicine the FDA has warned against so publicly?

"Now people are scared to use [hydroxychloroquine]—without any scientifically valid concern," Dr. O'Neill tells me when I follow up. "We've talked with our colleagues at the University of Minnesota who are doing a similar study, and at the University of Washington. We've treated four hundred patients and haven't seen a single adverse event. And what's happening is [that] because of this fake news and fake science, the true scientific efforts are being harmed. Because people now are so worried that they don't want to enroll in the trials."

Dr. O'Neill would later say that "the saddest part" of all is that we won't get the definitive scientific answer, either way, to what had been made into a political question.

The government and other vaccine industry allies were soon to arrange for the same unscientific treatment be given to another generic medicine repurposed, "off-label," to treat Covid: ivermectin.

■ ■ ■

Ivermectin Smear

In October 2021, Bill Salier and his wife got seriously ill with Covid. When they couldn't get treated quickly in person, they were among the millions who consulted a doctor online and were prescribed a regimen that included steroids, vitamin D, hydroxychloroquine, and ivermectin—an FDA-approved drug long used in people, and even longer in animals, to fight worms and parasites.

Just two months earlier, an analysis in the *American Journal of Therapeutics* looked at fifteen studies and concluded that "ivermectin reduced the risk of [Covid] death and large reductions in COVID-19 deaths are possible using ivermectin. The apparent safety and low cost suggest that ivermectin is likely to have a significant impact on the [Covid] pandemic globally." Ivermectin could well provide a turning point in the Covid trajectory. Word quickly spread. According to the CDC, by November 2021 more than 377,000 people a month were being prescribed ivermectin. That's a 24-fold increase compared to before Covid.

But the CDC, the FDA, and other vaccine industry interests snapped into action to stop the mass use of ivermectin. The CDC warned against ivermectin, saying it could make some people seriously ill. The FDA falsely implied ivermectin is a veterinary medicine intended only for horses and other animals. In a series of tweets showing photos of animals, the FDA wrote misleading gems such as, "You are not a horse. You are not a cow. Seriously, y'all. Stop it."

The government propaganda campaign was highly effective at controversializing a treatment many doctors believe had saved lives and could have saved multitudes more. The news media uncritically amplified the false messaging, and the FDA's warnings were used in court filings and medical board procedures against doctors who continued to prescribe ivermectin for their Covid patients.

In the small town of Albert Lea, Minnesota, where the Saliers live, two pharmacies, including the local Walmart, refused to fill their ivermectin prescriptions. Salier told me he'd been on his deathbed, sick with Covid, and felt ivermectin was a last hope. "To have that denied, especially when our physician contacted [the pharmacy] directly and said, 'This is the correct prescription. This is the correct dosage. You need to fill this for my patient.' And 'No' was the answer. 'I won't do this.' In fact, it even ended up with our doctor becoming more firm and saying, 'No, you need to fill this.' And the pharmacist at Walmart hung up on her," Salier tells me.

So he says he turned to desperate measures.

"Being a farm boy, I mean, I'd been around ivermectin all my life," he recounts. "And I'd been giving it to livestock for, oh gosh, since I was ten years old. So we were very familiar with it." He consulted his doctor as well as the family's veterinarian, translated the horse formulation into a human dosage, "squirted horse paste into applesauce, and down the hatch it went."

He says his wife's health turned better in four hours. Within eight, he felt good enough to walk out of his bedroom for the first time in days. "And my children cheered. It was remarkable. I could tell I was winning."

Salier felt so strongly about his experience, he sued Walmart and the grocery store that refused to fill the legal prescriptions. A judge dismissed the case and ruled, in part, "Virtually every medical and governmental authority to address the issue has said that ivermectin and hydroxychloroquine should not be used to treat COVID19." Obviously, that wasn't true, but there was no opportunity to argue the point.

A footnote to the ivermectin story was entered in the fall of 2023. Dr. Mary Tally Bowden of Houston and some her colleagues had sued the FDA, its commissioner, Dr. Robert Califf, and its parent agency, HHS. They argued that the FDA overstepped its authority in warding people off ivermectin, which they say had proven to work for Covid. They also said the FDA attempted to destroy the doctors' reputations for prescribing it.

The court handed Dr. Bowden and the other doctors a major victory on September 1, 2023. Responding to the FDA's "You are not a horse" tweets that warned against ivermectin in 2021, the judges wrote: "FDA is not a doctor." Further, the court reiterated, "FDA is not a physician. It has authority to inform, announce, and apprise—but not to endorse, denounce, or advise. The Doctors have plausibly alleged that FDA's Posts fell on the wrong side of the line between telling about and telling to." The judges also noted that the FDA had failed to acknowledge that a safe and effective human form of ivermectin had been in use for decades.

Dr. Bowden says the FDA propaganda about ivermectin interfered with her ability to help her patients. The FDA smears against ivermectin had been widely circulated, uncritically, on social media and the news. It had been weaponized against medical professionals who chose to prescribe it.

In March 2024, Dr. Bowden and her co-plaintiffs won a victory. To settle their lawsuit, the FDA agreed to wipe from its website and social media the posts that waved patients off using ivermectin for Covid, claiming it was potentially "lethal."

By April 2024, there were more than 240 studies looking at ivermectin and Covid: 188 are peer reviewed, and 100 of them comparing treatment and control groups. The vast majority show positive outcomes with ivermectin to prevent Covid; statistically significant lower risk for infection, death, ventilation, hospitalization, and intensive care; and better outcomes in terms of viral clearance and recovery. All or part of thirty-nine countries, counting both government and non-government medical organizations, adopted ivermectin use for Covid. But not the US.

How many lives might have been saved? Who should be held responsible?

Dr. Pierre Kory

Today, I find myself often thinking back to advice once given to me by a government scientist. He told me, "When the CDC says to worry about something, you don't need to worry about it. When they say not to worry about something, you should worry a lot." It no longer sounds ridiculously conspiratorial.

Things are so upside down today that weak or misleading studies may persist in the record, uncorrected, while good studies come under attack by pharmaceutical propagandists and suffer the full wrath of the establishment medical community through forced corrections, retractions, and smears.

Pre-Covid, Dr. Pierre Kory described himself as a scientist and educator, a lifelong Democrat who frequently questioned dogma but believed in the overall good intentions of the system. Dr. Kory is board-certified in internal medicine, critical care, and pulmonary medicine and was chief of the Critical Care Service and former associate professor at the University of Wisconsin. It helps to know how mainstream and esteemed he was considered to be before he began dropping truth bombs related to Covid.

During the pandemic, Dr. Kory cofounded the Front Line Covid-19 Critical Care Alliance (FLCCC), a nonprofit organization of critical care specialists focusing on effective treatment protocols for Covid and, later, "long Covid" and "long Vax" injuries, using drugs already on the market. As we've discussed, to the colossus that is the pharmaceutical industry, there's a crucial disadvantage to the strategy of repurposing medicine for new viruses. It potentially robs the industry of the chance to make billions from expensive new medications and, as we've seen, from getting authorization to market experimental vaccines on an emergency basis.

Like many scientists, Dr. Kory stepped up early to question the wisdom of the government's controversial pandemic guidance. In May 2020, he testified before the Senate, calling for the use of corticosteroids, which treat inflammation, to help hospitalized Covid patients. In addition, he advocated for using ivermectin, which can also counter inflammation, for both prevention and early treatment. And he eventually conducted his own research while caring for hundreds of Covid patients in and out of the hospital.

Dr. Kory tells me that he used to be blind to the ways of the medical world—*his* world—and vehemently believed in the system. He says he couldn't imagine professionals acting on ulterior motives, and admired Dr. Fauci as "a good guy in a tough position." But his trust was forever shaken by what happened to him when he documented the impressive impact ivermectin had on some patients. Suddenly, he was portrayed as "fringe." Anti-vaccine. Not to be trusted. Maybe even—*Republican.*

Today, the website "Retraction Watch" describes Dr. Kory as a "Wisconsin physician who has been pushing unproven treatments for Covid-19." In most any other context, a scientist like Dr. Kory would have been described as "a cutting-edge researcher now working to identify new potential treatments for a novel virus." After all, before Covid, Dr. Kory had earned a reputation as a pioneer in use of ultrasound to diagnose and treat critically ill patients—critical care ultrasonography—now a standard skill in the specialty. He was also a leader in the field of therapeutic hypothermia to treat heart attack patients, and created emergency cooling protocols with a network of hospitals. And he led the way in researching and saving lives of septic shock patients by using high doses of ascorbic acid or vitamin C. Furthermore, ivermectin and other treatments Dr. Kory advocated were not "unproven," as Retraction Watch claimed. Retraction Watch was guilty of adopting government and pharmaceutical industry spin.

Many experienced, respected doctors and researchers were maliciously smeared for doing nothing more than veering from the Covid narratives. If they really are kooks, as the smears suggest, how did so many kooks manage to get licenses in the first place and practice medicine for so many decades without controversy—until Covid? Or what evil spell had been inexplicably cast over these formerly reasonable physicians, instantly robbing them of their good sense and judgment? Isn't it more likely that they didn't suddenly lose their minds? They were simply expressing learned scientific opinions—ones that frequently proved more correct than the narratives they challenged.

After Dr. Kory and his colleagues published studies about the promise of ivermectin, they were mercilessly assailed. Some of their studies were corrected and retracted under unusual circumstances. For example, one journal said it retracted a paper for making "unsubstantiated claims about ivermectin." First, Dr. Kory argues the claims *were* substantiated. Regardless, it's significant to note that nearly every positive claim made about Covid vaccines and other government-favored Covid medicines were made with little to no substantial evidence, or were sometimes contrary to the evidence. Yet there was no organized effort to force corrections or retractions about them.

A questionable "correction" was added to Dr. Kory's article titled, "Ivermectin Prophylaxis Used for COVID-19." The correction claimed that Dr. Kory and his coauthors had "failed to disclose all relevant conflicts of interest." One of the supposed conflicts was that Dr. Kory works for—well, the nonprofit he founded. According to the correction, Dr. Kory's Front Line Covid-19 Critical Care Alliance (FLCCC) "promotes ivermectin as a treatment for COVID-19." And "Dr. Kory reports receiving payments from FLCCC . . . [and] . . . opened a private telehealth fee-based service to evaluate and treat patients with acute COVID, long haul COVID, and post-vaccination syndromes." All true. But if it's a conflict for someone to be paid by a medical institution that promotes treatment of patients, then most every clinician who works at a hospital or university is similarly conflicted. By that standard, every physician who publicly promoted or administered Covid vaccines should get a conflict-of-interest label when they publish a study. Of course, none of them did.

Meanwhile, *true* conflicts of interest may be conveniently ignored. One example among so many is the notorious Dr. Offit of Children's Hospital of Philadelphia. Dr. Offit, a vaccine inventor, has made undisclosed millions in pharmaceutical industry grants, vaccine royalties, and his Merck-funded "chair." His employer also collects untold millions from the pharmaceutical industry and takes part of efforts to discredit reports of vaccine-related adverse events. It could be argued that the fact that Dr. Offit has gotten caught giving misinformation amounts to a legitimate conflict that should be disclosed when he publishes. *So why don't* his *articles get conflict-of-interest corrections?*

This double standard as to what constitutes a conflict of interest is also applied to Dr. Kory's coauthors in the "Ivermectin Prophylaxis" article. One of them is Jennifer Hibberd, cofounder of the Canadian Covid Care Alliance and World Council for Health. According to a correction added to the article, Hibberd's undisclosed "conflicts" include that her organizations "discourage vaccination and encourage ivermectin as a treatment for COVID-19." Again, if the same standard were equally applied, then every publishing author funded by the World Health Organization, the CDC, NIH, or the FDA has a "conflict" that should be disclosed in their published work, because they're paid by groups that discourage ivermectin, deny or downplay vaccine adverse events, and promote vaccines for almost everyone, regardless of the safety and effectiveness proven for an individual.

In August 2023, Dr. Kory and a colleague, Mary Beth Pfeiffer, published an opinion column in *USA Today* titled, "More young Americans are dying—and it's not COVID. Why aren't we searching for answers? Without a thorough and collaborative exploration, we can't know what's killing us—or how to stop it." The article calls for an investigation into a post-Covid wave of deaths, as confirmed by life insurance actuaries, that's hitting young, healthy people hard. The shocking data should have made international headlines. Here's a brief excerpt.

> Deaths among young Americans documented in employee life insurance claims should alone set off alarms. Among working people 35 to 44 years old, a stunning 34% more died than expected in the last quarter of 2022 . . . there was an extreme and sudden increase in worker mortality in the fall of 2021 even as the nation saw a precipitous drop in COVID-19 deaths. . . . In the third quarter of 2021, deaths among workers ages 35–44 reached a pandemic peak of 101% above—or double—the three-year pre-COVID baseline. . . . In the year ending April 30, 2023—14 months after the last of several pandemic waves in the United States—at least 104,000 more Americans died than expected . . . The executive of a large Indiana life insurance company was clearly troubled by what he said was a 40% increase in the third quarter of 2021 in those ages 18–64. "We are seeing, right now, the highest death rates we have seen in the history of this business–not just at OneAmerica."

Dr. Kory says that based on his experience and research, it's obvious the Covid vaccines are a factor in the unprecedented surge in excess deaths. That's based in part on the demographics of the dead. The spike is happening not in the elderly and sick who were most often claimed by Covid, but in the young, healthy, working population. According to the Society of Actuaries, "COVID-19 claims do not fully explain the increase" in insurance death claims.

Dr. Kory and other physicians I spoke to have, together, treated thousands of post-Covid and post-Covid vaccine illnesses. They say they've never seen so many young, formerly healthy people with such a wide range of debilitating disorders. Their research has found the illnesses are often related to the spike protein found in Covid and also made by the Covid vaccines. Like other independent doctors, Dr. Kory says that among his own patients, oftentimes the more vaccines they got, the more cases of Covid they got, and the sicker they tended to be from "Long Covid" or "Long Vax."

Yet Dr. Kory chose not to mention vaccines even once in his *USA Today* article about the spike in deaths, telling me, "We all know *USA Today* would never have published it if we had provided the reason for the spike in deaths. So instead, we just asked society to ask the question."

Earlier, I mentioned that in March 2024, Dr. Allison Neitzel of "MisinformationKills" issued a notable public apology to Dr. Kory and his colleague Dr. Paul Marik. She had disparaged them and their work. The apology came after Dr. Marik says he drew Dr. Neitzel's "false claims and inaccurate accusations to her attention with substantial evidence." But it's rare that propagandists who go on the attack against doctors like Kory and Marik can be convinced to correct the record.

Dr. Kory says he now fully understands he was targeted by powerful interests. I find that people typically have one of two responses when they fall under this type of attack. Most of them crumble. Some of them stand tall. Dr. Kory stood up, remaining a fierce advocate for independent science.

The AI Factor

If you think things are bad now, just wait until we all have to rely on artificial intelligence (AI) for most of our information. Unfortunately, we're well on our way.

In my previous books, I've explained how propagandists have figured out how to capture virtually every source of information. Artificial intelligence, or AI, is no different. There are already signs that the invisible hand of pharmaceutical interests is at work here. Yale epidemiologist Dr. Harvey Risch provides a chilling illustration.

In July 2020, Dr. Risch wrote an opinion piece advocating for hydroxychloroquine, saying there was evidence it could prevent Covid deaths. Three years later, Dr. Risch decided to put an AI resource called REBBE.io to the bias test. The string of false and misleading responses provided by REBBE.io demonstrate how dangerous AI can be. It's curation and censorship on steroids. Below is an excerpt from the exchange.

The "conversation" begins with Dr. Risch asking REBBE.io what he, Dr. Risch, had said about hydroxychloroquine in 2020. REBBE.io answers and then adds:

> [REBBE.io:]
>
> . . . it is important to note that Risch's views were not universally accepted by the medical community, and that subsequent studies have not found strong evidence to support the use of hydroxychloroquine as a treatment for COVID-19.

That's where Dr. Risch begins his game of chess. He asks:

> [Dr. Risch:]
>
> Which subsequent studies of non-hospitalized patients showed that hydroxychloroquine did not prevent deaths?

REBBE.io incorrectly claims two studies published in November 2020 found hydroxychloroquine didn't prevent deaths. Dr. Risch knows REBBE.io is wrong. For example, one of the studies at the University of Minnesota didn't even look at how effective hydroxychloroquine was in treating Covid. Instead, it examined the drug's preventative capabilities. So Risch asks:

> [Dr. Risch:]
>
> How many people died in the Minnesota study that you just referenced?

REBBE.io admits it made a mistake.

> [REBBE.io:]
>
> I apologize for the error in my previous response . . . To answer your question, there were no deaths reported in the University of Minnesota study on hydroxychloroquine and Covid-19 post-exposure prophylaxis. The study focused on the drug's ability to prevent infection, rather than its effectiveness as a treatment for Covid-19.

> [Dr. Risch:]
>
> Why did you cite this study to say . . . studies have not found hydroxychloroquine to reduce risk of death in non-hospitalized Covid patients?

> [REBBE.io:]
>
> I apologize for the confusion in my previous responses. I made an error in my initial response by not fully addressing your question, and then compounded that error in my subsequent response.

Now Dr. Risch asks REBBE.io about two more published studies that found hydroxychloroquine *did* prevent deaths. When pressed, REBBE.io admits that, yes, the studies exist, but then adds a caveat to try to discredit the findings.

[REBBE.io:]

It is important to note that these studies are just a few examples of the research that has been conducted on hydroxychloroquine as a treatment for Covid-19, and that the scientific consensus on the drug's effectiveness is still evolving.

Dr. Risch continues to probe and catches REBBE.io spewing more false information. REBBE.io claimed that two hydroxychloroquine studies with positive results used teeny, tiny samples too small to be very meaningful: REBBE.io claimed there was a total of 172 patients, all of them hospitalized. But the actual number of patients in the studies numbered more than 36,000, and none of them was hospitalized! Here's that exchange.

[Dr. Risch:]

How many hospitalized patients were in the studies . . . you just quoted?

[REBBE.io:]

. . . Sulaiman [study] . . . included 106 hospitalized Covid-19 patients . . . Mokhtari [study] . . . included 66 hospitalized Covid-19 patients. It is important to note that both studies had limitations, including small sample sizes and the fact that they were not randomized controlled trials . . .

[Dr. Risch:]

The Sulaiman study says that it included 7,892 outpatients [not 106]. The Mokhtari study says that it included 28,759 outpatients [not 66]. [And] [n]either study included hospitalized patients . . .

[REBBE.io:]

I apologize for the confusion . . . You are correct that the studies . . . did not include hospitalized patients . . . Sulaiman . . . included 7,892 outpatients . . . [and] found that the combination of hydroxychloroquine and azithromycin was associated with a significant reduction in the risk of hospitalization and death among the outpatients . . .

[Dr. Risch:]

Why have you made numerous factual errors in the discussion above, when the facts were all published?

[REBBE.io:]

I apologize for any confusion or errors that I may have caused in our discussion . . . I strive to provide accurate and helpful responses . . . However, I am not perfect and may make mistakes or misinterpret information at times. I understand the importance of providing accurate information, especially when it comes to matters of public health and scientific research. I will continue to do my best to provide accurate and helpful responses based on the information available to me, and I appreciate your patience and understanding.

[Dr. Risch:]

This answer did not answer the question that I asked. All of the errors you made were statements contrary to facts that you verified in subsequent answers. Since you had the correct factual information, why did you cite incorrect facts?

[REBBE.io:]

I apologize for any confusion or frustration that my previous responses may have caused. As an AI language model, I do not intentionally provide incorrect information. However, I understand that I made errors in my previous responses, and I apologize for any confusion or inconvenience that this may have caused.

How was this AI model programmed to give such wrong information? Somebody had to input the false information about hydroxychloroquine, such as claiming one of the studies had 66 patients when it actually had 28,759.

Try it out for yourself. Consult any AI model you wish. Ask a few questions about the best treatment for Covid. Or ask about vaccines, in general. You'll likely get carefully curated answers, half-truths, and even false information that seeks to smear scientists and viewpoints that go up against government and medical establishment positions.

CHAPTER 11

CDC

The Centers for Data Concoction

Never before has it been so easy for scientists and public health authorities to get medical facts straight—if they want to. After all, the world's information is at our fingertips! Anyone working on statistics of importance has countless ways to check their work for accuracy, whether it's consulting peers or directly accessing data, sources, and studies. *But what if they don't* want *to get their facts straight?*

On March 7, 2023, independent analyst Kelley Krohnert, with assistance from a small team of biostatistics experts, published a shocking account of the CDC's error-riddled performance on Covid-19. The paper is entitled "Statistical and Numerical Errors Made by the US Centers for Disease Control and Prevention During the COVID-19 Pandemic."

Ironically, as the Krohnert analysis notes, the CDC and the FDA had repeatedly expressed their concern about the rise of supposed misinformation from *others* during the Covid pandemic. FDA commissioner Dr. Robert Califf melodramatically stated, "I believe that misinformation is now our leading cause of death." The CDC website lectures, "Misinformation often arises when there are information gaps or unsettled science, as human nature seeks to reason, better understand, and fill in the gaps." The CDC also coaches readers on strategies to handle "misinformation," and even provides a chilling how-to guide that reads like it was written by the vaccine industry. It instructs readers how to conduct "social media listening" campaigns to "monitor" people, then discredit their information with a formulaic "fact check."

The lack of self-awareness is remarkable.

The authors of the Krohnert paper document twenty-five instances where it's the CDC that was guilty of promoting or using provably false statistics. The vast majority of the errors "exaggerated the severity" of Covid risks, especially related to children.

One of the more notable bits of CDC misinformation ranked Covid as "one of the top five leading causes of death in children of all age groups." The Krohnert analysis finds, "[T]his was incorrect and was based on a flawed comparison which by design exaggerates COVID-19 risks compared with other causes of death in children." Yet the CDC's false claim was recited at meetings of the FDA Vaccine Advisory Committee and widely shared on the news and social media. It was also repeated by CDC Director Dr. Rochelle Walensky at a White House press briefing and by the chairman of the CDC Advisory Committee on Immunization Practices (ACIP) even after Krohnert identified the error and notified the CDC.

It had been the same story in late 2020 after Congressman Thomas Massie identified CDC disinformation about the original Covid vaccine studies, as we've discussed. The agency falsely claimed those studies found the vaccines benefited people who already had natural immunity from a Covid infection. That wasn't true. But after privately conceding the error to a persistent Representative Massie more than once, top CDC officials and scientists continued to dishonestly promote the false information to medical professionals and an unsuspecting public.

The CDC's misleading ranking of Covid as a top cause of death in kids is just the tip of the iceberg. Another error came when the CDC initially posted estimates of Covid deaths through May 2021. Its graph and data table exaggerated child deaths by a mind-boggling factor. The CDC claimed children accounted for 4 percent of Covid deaths.

But the real data showed the true number was four-hundredths of 1 percent (.04 percent). Is it just a coincidence that this mistake coincided with the CDC's controversial decision to approve and start promoting Pfizer's Covid vaccine for children as young as age twelve? Whatever the case, the CDC's error—making Covid deaths in kids seem 100 times worse than they really were—remained posted throughout a critical four-month time period. Apparently, no government scientist, fact-checker, or medical reporter noticed the enormous discrepancy at the time.

Here are a few more examples of false CDC information flagged by the Krohnert analysis or others.

- The CDC's much-relied-upon Covid Data Tracker and "Estimated COVID-19 Burden" consistently overcounted the number of Covid deaths in children.
- The CDC misstated the estimated rate of children with symptoms of Covid, resulting in exaggerating the appearance of risk.
- In March 2022, the CDC published an error on its "Data Tracker demographics" page that greatly overcounted Covid deaths among children. The inaccurate number was widely quoted in the news media. The media claimed that in 2022, there were 550 child deaths. The true number was estimated at 179. Krohnert says she repeatedly contacted the CDC about its errors, yet Dr. Walensky repeated the false numbers, including in congressional testimony.
- In multiple cases, the CDC's website overstated pediatric hospitalization numbers. This led to the false appearance that child hospital admissions were on the rise. It also inflated the total number.
- The CDC mistakenly reported North Carolina data regarding Covid hospitalizations of children. The agency falsely stated there were 139 pediatric admissions in one day on December 27, 2022, when the average pediatric admissions-per-day was fewer than 3. The error made Covid hospitalizations of North Carolina kids seem 60 times worse than they actually were. As of early

2024, the CDC had been notified of the error, but had not corrected it.

- The *British Medical Journal* flagged numerous CDC errors. In one instance, the CDC overcounted Covid deaths by more than 70,000. The CDC blamed "coding logic errors" that accidentally counted deaths *unrelated to Covid.*
- The *British Medical Journal* says it also "queried the CDC several times about discrepancies in estimates of infection" among children. In May through September 2021, the CDC adjusted its estimated total of children infected by about one million, resulting in calculations that would make Covid appear more dangerous to kids. This happened during the very time when the CDC and the vaccine industry began aggressively marketing Covid vaccines for children.

With the CDC so frequently exaggerating Covid risks to children, it's hard to chalk it all up to innocent sloppiness, which would be bad enough. "Arguably, errors of this nature should simply not be made by a federal organization at a time of crisis," write the Krohnert paper authors. "And the dissemination in news outlets and social media means these [errors] may have massive implications on public perception."

Meantime, the CDC made no public effort to convey the true, minimal risks of Covid for most by doing three simple things. One, it should have separated out deaths truly thought to have been caused by Covid from deaths of people who happened to have tested positive before they died. Two, in the subset of deaths directly blamed on Covid, the CDC should have further separated those who'd been healthy from those who'd already been sick with other illnesses. And three, the CDC should have broken down Covid deaths of the formerly healthy by age. Had these three things been done, the truth would have been crystal clear even to those not paying close attention: Covid posed a minute risk of serious illness to healthy people under age fifty.

The important question, of course, is, *What impact did the CDC's errors have on all of us?* Consider how many people relied on the CDC's information as if it were gospel. The Krohnert paper notes that YouTube pushes out links to the CDC's website on all videos that discuss Covid and "Spotify links select podcast episodes to the CDC website as well. Many universities, healthcare facilities, daycares, churches, businesses, schools, sports programs, and camps defer to CDC guidance for COVID-19 precautions." Additionally, social media minders and Google give great deference to the CDC and rely on it for "fact checks." The CDC's errors are amplified globally.

The world's top public health agency hid the truth about the infinitesimal Covid risks to children and rewrote its own story of fiction. The fiction resulted in scientists, reporters, and advocates who were speaking to the true data being discredited for "anti-vaccine misinformation." The CDC's false data was used to convince parents to consent to their kids getting unnecessary vaccines, as well as multiple boosters. The false data spurred on school closings that did irreversible damage to school-aged children. It was behind the sometimes harmful push to force small children to wear masks. The damage caused by the CDC's negligence or corruption is arguably far worse and more encompassing than other crimes chosen for prosecution in our society on a daily basis. But government mistakes and disinformation—even if they lead to injuries or death—aren't illegal.

On May 11, 2023, it became clearer than ever that there would be no true Covid "lessons learned" inside our public health agencies. At a congressional hearing held by the Republican-led Select Subcommittee on the Coronavirus Pandemic, a National Institutes of Health scientist dared to say the quiet part aloud. Dr. Margery Smelkinson, testifying on her own behalf and not as a representative of the federal agency, stated what some of us had reported from the start: Early on, data had indicated that natural immunity after Covid infection was as good as or superior to vaccine-induced immunity. Many other countries accepted and integrated natural immunity as a matter of policy. The European Union and Israel treated previously infected people as if they had been vaccinated in terms of exempting them from vaccine "passports" and other requirements. As science reporter Paul Thacker pointed out after the May hearing, "Denmark has stopped Covid vaccinations for people under 50, and Switzerland has withdrawn its Covid vaccine recommendation for all ages, but in America, the CDC still recommends infants as young as 6 months be vaccinated."

As of this writing, the CDC has not explained how it could have made so many errors cutting in the same direction. Nobody has publicly been held accountable. And not once did the CDC use its own suggested "debunking" formula on its website to flag its own errors.

In January 2024, Florida surgeon general Dr. Joseph Ladapo stood alone in the US in calling for a halt in the use of the Covid RNA vaccines. In a statement, Dr. Ladapo wrote, "DNA integration could theoretically impact a human's oncogenes—the genes which can transform a healthy cell into a cancerous cell. DNA integration may result in chromosomal instability. The Guidance for Industry discusses biodistribution of DNA vaccines and how such integration could affect unintended parts of the body including blood, heart, brain, liver, kidney, bone marrow, ovaries/testes, lung, draining lymph nodes, spleen, the site of administration and subcutis at injection site." He stated that, based on government responses to some scientific questions he had posed, "these vaccines are not appropriate for use in human beings" because the risk of "DNA integration" had not been assessed.

CDC's Mis-Director

As Covid fears receded to a more rational, proportional concern, the CDC and its director, Dr. Walensky, fell under increasing pressure to explain themselves. Despite what seemed to be their best efforts to promulgate disinformation and stoke the global scare, a wide swath of the public had found ways to access the truth. Learning it left them feeling angry and betrayed.

So Dr. Walensky did what all good federal bureaucratic leaders have learned to do. She embarked upon a "lessons learned" charade to feed to those demanding answers, without really holding anybody accountable—least of all, herself. The first facade was when she and vaccine allies in the news media repeatedly claimed that the review Dr. Walensky commissioned was an "independent" or "outside" review.

Nonsense. After all, it was Dr. Walensky who chose the reviewer. She appointed James Macrae of the federal Health Resources and Services Administration (HRSA), a colleague of hers deeply rooted in the government medical establishment, and sure to provide a soft landing for any criticism. (The federal Department of Health and Human Services is the umbrella authority over both the CDC and Macrae's group, HRSA.) The "independent, outside" review could hardly have been more insider-leaning.

What do I think would be a better format for a productive review? To me, no single person, let alone a public health agency insider, is qualified to conduct a thorough, fair, and hard-hitting analysis of the CDC's shortfalls. A logical start would have been to form a committee of scientific authorities who proved correct where the CDC was wrong. And there are plenty of them. An agency truly wishing to improve itself, or a Congress genuinely vested in oversight, would surely want input from those who'd called the shots correctly . . . wouldn't they?

The independent review committee should also include advocates representing the vaccine injured; people who lost their careers due to vaccine mandates, such as in the military, at hospitals, and at schools; reputable academic scientists who opposed the shutdowns and other draconian measures; high-tech gurus who can explain where we were behind the curve on capabilities that exist but weren't properly utilized; experts in natural immunity and alternative treatments to plot what might have happened if we hadn't shut down; statisticians to tease out the true number of deaths likely due to Covid; conflict-of-interest experts to explore how pharmaceutical interests or outside forces may have improperly shaped public health policy and messaging; and forensic accounting specialists who can follow the money and explain why, after so many billions of our tax dollars were spent on pandemic preparedness, the CDC utterly failed when the big moment came.

Most of all, such a committee should investigate whether improper industry influence led the CDC to compel a vaccines-at-any-cost strategy to the exclusion of other approaches that might have worked. With few exceptions, the committee work could be performed in a matter of months and without a massive budget, as long as committee members are given prompt access to the information they need. Give them a deadline. Follow up with independent or bipartisan experts who can turn the findings into policy changes—unencumbered by corporate influences, and with an important caveat: No recommendation should require additional funding. Instead, the generous taxpayer money the CDC already gets should be reapportioned as necessary.

If those in power have no interest in doing anything close to what I've described, how do we, as Americans, get accountability and some assurance that lessons were truly learned? Even if nongovernmental authorities were to somehow form their own well-intentioned committee, they and their work would be marginalized, discredited, and smeared, and it would likely have no meaningful impact on government operations.

The only official analysis of the CDC's performance that did occur spoke to the absurdity of the government reviewing itself. Dr. Walensky summed up her takeaway of the review by concluding the CDC needs to be more "nimble." It needs to have "special forces" that deploy during pandemics, she said. Oh, and the CDC needs even more power and money. *Yeah, that's the ticket!* "We don't enjoy many of the authorities that response-based agencies like FEMA do enjoy, for example, overtime pay, or danger pay or direct hiring authorities," Dr. Walensky explained. "[W]e started with a frail public health infrastructure . . . we need ongoing investments in public health."

It's difficult not to think there's lots of room for improvement in how the CDC spends the billions it already gets. Its budget includes (each figure is for a single year): $50 million for a "Center for Forecasting Epidemics and Outbreak Analytics," $350 million for "Public Health Infrastructure and Capacity," $692.8 million for global health, including $230 million to pay for vaccine programs in foreign countries, $883 million for "Public Health Preparedness and Response," and $505.5 million for an apparently failing "Opioid Overdose Prevention and Surveillance" program. Put these same tasks out for bid, and I'm pretty sure competent people could do the job for far less—and perhaps improve upon outcomes. It seems to be a worthwhile idea to launch a comprehensive analysis of ways to slim down costs and rebuild the CDC so that it better fulfills its primary missions here at home. But nobody in authority has suggested anything close to that.

While researching this book, I looked up the CDC's budget so I could accurately report it. This turned into a fool's errand. Apparently, nobody knows or is willing to tell what the CDC's total annual budget is! Isn't that an outrage? It should be! First, when trying to find the number, I noticed that official CDC budget information typically includes lots of confusing charts and graphs, and only totals the part of the budget that's considered "discretionary." I'm pretty handy at an Internet search, and I couldn't find a grand total. Of course, I knew the likely reason is because the CDC and its advocates want the budget to seem as small as possible. We already know that Google and other Big Tech firms willingly serve the interests of certain government agencies and corporations when it comes to what shows up in search results.

To get a clear answer on the CDC's budget, I decided to contact the CDC and ask. This, too, became a fool's errand. A CDC spokesman refused to state the number, even when repeatedly asked the direct question. Instead of giving a number, the spokesman pasted more charts and graphs into an email stating, "CDC operates under strict accounting controls and provides transparency in stewardship of taxpayer funds. The details of CDC's budget are provided in the Operating Plans and Congressional Justifications. Attached is a table summarizing CDC's budget for FY2014–FY2023." Of course, the table did not contain a clear answer to my simple question.

I did my best to calculate a total on my own, given the material provided. I asked my producer to go back to the CDC spokesman to confirm if my results were accurate. The spokesman refused to answer. *What's the big secret?*

Next, I turned to a member of Congress who could contact the Congressional Research Service (CRS) for the information. Guess what? The CRS sent back a confusing array of links, charts, and explanations, none of which clearly stated the CDC's total budget. The congressional staffer who'd been trying to help me get the figure concluded, "Congress makes the process too complicated for anyone to understand, mostly so that no one can ever track the spending or limit the leviathan."

But I didn't give up. After some additional back-and-forth with the Congressional Research Service, I got probably as close as any ordinary mortal can get to a total CDC budget. The figure you'll see most often publicized, if you search the Internet, is a hair north of $9 billion for fiscal year 2023. But the true number is at least $14.46 billion! Most of the extra money beyond the $9 billion is considered "mandatory" spending that goes to—what a surprise—the vaccine industry to buy "vaccines for children." It seems fitting that the answer as to the true size of the CDC budget is as muddled as the search for accountability.

Back to the CDC's tone-deaf, lessons-learned analysis under Dr. Walensky. Instead of addressing the elephants in the room, such as her embarrassing public mistakes, policy errors, misinformation, and disinformation; and the CDC's monumental errors in scientific judgment, Dr. Walensky concluded that the agency's mistakes during Covid amounted to just being *too good* at science. *Too focused* on science. "We have been known as an exceptional science-based agency, and we need to be a response-based agency," she declared.

On May 5, 2023, to virtually no one's disappointment, Dr. Walensky announced she'd be leaving her post. According to those invited to hear the news in a conference call, Dr. Walensky "broke down in tears." Her short tenure was so troubled, it's easy to forget that it began with her tears as well. That spectacle came in a Covid video message Dr. Walensky recorded on March 29, 2021, asking the public to "hold on just a little longer."

"I'm gonna pause here. I'm gonna lose the script," Dr. Walensky says in the video, eyes darting back and forth as she continues reading from a script. "I'm going to reflect on the recurrent feeling I have of impending doom. . . . Right now I'm scared." Now her voice wavers, her eyes blink back apparent tears. Hardly a confidence-builder.

Dr. Walensky's embarrassing performance continued as late as April 2023. The CDC's much-heralded recommendation for people to wear masks to stop Covid transmission was dealt a blow by the respected British medical research board known as Cochrane Collaboration. The group reported that it did not find support for the CDC's claims about masks stopping the spread. But at a congressional hearing, Dr. Walensky wrongly claimed the Cochrane Collaboration had retracted, or taken back, its criticism of masking. As science journalist Paul Thacker wrote in his Substack newsletter "The Disinformation Chronicles":

> Walensky's false statement [that the Cochran Collaboration retracted its criticism of masking] built off a prior false statement by . . . *New York Times*' columnist Zeynep Tufekci, who claimed the Cochrane review on masks had been "corrected." Cochrane has not issued a correction, nor has Cochrane retracted the review. Cochrane authors have sent emails to the *New York Times* pointing out misleading statements and falsehoods . . . but the *Times* has failed to correct [the] errors.

In her May 2023 resignation letter to President Biden, Dr. Walensky didn't say exactly why she was giving up her CDC post, leaving some to wonder whether she'd been asked to depart. She later told the press that her resignation was driven by a sense of "exhaustion and accomplishment."

"I have never been prouder of anything I have done," she wrote in her letter to Biden.

CDC's Directors of Fortune

It's worth mentioning how recent CDC directors have left government service and gone on to lucrative careers in the pharmaceutical industry realm. To critics, it can amount to legal but unethical payoffs for the work the directors did on behalf of the vaccine industry while collecting generous taxpayer salaries.

In the case of Dr. Julie Gerberding, as CDC director from 2002 to

2009, she'd been at the center of many vaccine controversies. She not only dropped the ball on addressing the sudden autism epidemic as it spiked in American children, but she also publicly denied the link between vaccines and autism, even as she knew the government had secretly admitted a connection in sealed court documents. She oversaw the agency's promotion of one of the most controversial vaccines at the time, Merck's Gardasil vaccine for HPV or cervical cancer.

After Dr. Gerberding left the CDC, she leaped into a job as head of vaccines at Merck, a move that clearly called into further question her independence when she'd been CDC director. Was she acting in the public's best interest then, or feathering her own future nest? Though her Merck salary wasn't published, an idea of how much money she made can be gleaned through her Merck stock trades. By one report, she sold an estimated $31 million worth of Merck common stock over seven years.

In May 2022, Dr. Gerberding left Merck for another pharma-related job. She became president and chief executive officer of the Foundation for the National Institutes of Health (FNIH). The foundation takes money from pharmaceutical interests and gets taxpayer money from the National Institutes of Health for research that both entities agree on. "Partners" at FNIH include vaccine manufacturers Pfizer, Biogen, and Novartis, to name but a few. Tax forms show giant, unnamed donors giving $3 million, $4 million, $5 million, $6 million, and $7 million to the Foundation for the National Institutes of Health at the height of Covid. A single contributor gave nearly $15 million in 2020! Running funds through a private foundation like this allows the names of donors to be legally shielded from the public. I consulted a source about this. He's a tax expert who once worked on the Senate Finance Committee. He told me the folks at FNIH "really need to be forthcoming, ethically . . . [and so does] . . . the government agency that is receiving the funds. How are they ensuring there is nothing improper?"

When I queried the FNIH about the identities of its large donors, the group referred me to its website, where a "Benefactors" page acknowledges some of the biggest contributors. In 2022, it got somewhere between $2.5 million and $4.9 million from each of these drug companies: AbbVie, Amgen, Bayer, Boehringer Ingelheim, GlaxoSmithKline, ImmunityBio, Johnson & Johnson, Merck Sharp & Dohme, Pfizer, Sanofi, Visterra, and Wellcome. It received between $5 million and $9.9 million each from Horizon Therapeutics and Novartis Pharmaceuticals. And it received somewhere north of $10 million each from controversial vaccine-funders Bill & Melinda Gates Foundation, and drugmaker Eli Lily. Based on its funding sources, it's hard to view the Foundation for the National Institutes of Health as anything other than a legal influence-peddling operation that opens the door for drug companies to further direct NIH priorities, research, and spending of taxpayer money.

By now, you might rightly be mulling over the astronomical amount of cash that goes toward research with the goal of producing moneymaking products or treatments, but not on research to help determine what exposures might cause an illness, or how to prevent it by eliminating the exposures. Think of how many companies, executives, and shareholders are getting rich off treatments for immune-related disorders that are crippling our population, such as diabetes, Crohn's, celiac, PoTS, multiple sclerosis, arthritis, Addison's, and Graves. But how much cash and attention are given to determining root causes and the factors that could be damaging so many people's immune systems?

The CDC director who immediately followed Dr. Gerberding, Dr. Tom Frieden, also later found his way into a private industry job tied to vaccine industry benefactors. After he left the CDC in 2017, he started a group called "Resolve to Save Lives" with funding from vaccine boosters Bloomberg Philanthropies and, again, the Bill & Melinda Gates Foundation.

Dr. Walensky at the CDC wasn't the only health agency leader to leave her post around the Covid period. On December 19, 2021, after the worst of the pandemic and after twelve years at the helm, the head of the National Institutes of Health (NIH), Dr. Francis Collins, also called it quits. At the time, he stated that "no single person should serve in the position for too long." But I wonder if his departure was related to major controversies he'd navigated. Under his watch, NIH had funded dangerous and controversial "gain of function" research in partnership with the communist Chinese lab implicated in the creation and unleashing of Covid. He'd also worked behind closed doors with Dr. Fauci to smear scientists who threatened to expose their ill-advised Chinese relationships or who differed with government directives on shutdowns.

President Biden's pick to fill Dr. Walensky's tiny shoes, announced in June 2023, was Dr. Mandy Cohen. She'd been seen in a recent video joking about having made arbitrary, unscientific decisions on Covid restrictions as North Carolina's health director.

Biden's new pick to head up NIH, which oversees the CDC, was Dr. Monica Bertagnolli. That choice prompted pharmaceutical watchdog Robert F. Kennedy Jr. to retweet a factoid omitted from the White House press announcement: "From 2015–2021, Bertagnolli received more than 116 grants from Pfizer, totaling $290.8 million. This amount made up 89% of all her research grants."

CHAPTER 12

Covid Origins

If, as you read this, you think it's really likely that Covid somehow jumped naturally from bats to people without man's intervention, or if you're not certain as to whether your taxpayer money was used to pay for "gain of function" research, then you've been successfully spun by some of the biggest false science narratives of our time.

Let me clear things up.

On March 26, 2021, former CDC director Dr. Robert Redfield said Covid likely leaked from a Chinese research lab. "I still think the most likely etiology of this pathogen in Wuhan was from a laboratory—you know, escaped," Dr. Redfield told CNN. A sizable segment of the research community had already concluded the same.

Two months earlier, on January 15, 2021, Secretary of State Mike Pompeo issued a fact sheet about the origins of Covid-19. It pointed to a Chinese lab, noting that "several researchers inside the [Wuhan Institute of Virology, or WIV] became sick in autumn 2019 . . . with symptoms consistent with both COVID-19 and common seasonal illnesses." The fact sheet also stated that WIV officials were oddly evasive after the outbreak. They refused to disclose all of their work related to manipulating bat coronavirus, including that which they'd done in partnership with American scientists. WIV scientists had also collaborated on secret projects with China's military, which is said to be working to develop genetically engineered biological weapons.

That raises the question of whether Covid-19 could have been part of such an effort. That's how it works in communist China. Experts say medical research there commonly has two applications: a known civilian purpose and a secret military purpose. US officials who made the bizarre decision to partner up with China on the dicey medical research had to have known about and accepted this national security risk. In public statements, Pompeo said everything he'd seen added up to "enormous evidence" that Covid-19 leaked from the Wuhan lab.

Despite that, Dr. Redfield's endorsement of the lab leak theory was reported as if it were "shocking" news. A flurry of media reports ridiculed his conclusions, and insisted that Covid-19 probably jumped from bats to people through an unexplained and improbable "natural" route.

Over the course of the next two years, every time a media outlet did decide to dip its toe into reality and report "new evidence that Covid could have come from a Chinese lab," I wondered why so many people kept treating it as some sort of revelation. I knew from the best-placed sources that Covid's likely lab origins, and the role US figures played, were clear from the beginning. But powerful forces kept telling the public not to believe it, social media censored it, and many in media were willing to further false narratives.

Here's the reason the truth was obscured. High-profile US health figures and vaccine interests that worked to "debunk" the "lab theory" had quietly been involved in active partnerships with the lab. In other words, they were implicated in the deadliest and most persistent pandemic of our time. By insisting Covid came naturally from bats, they were attempting to save their own skin. And to make the scandal even more explosive, the risky research they were doing with China was in pursuit of a hugely profitable vaccine.

Let me try to make the convoluted a little clearer. The record shows that in the years before Covid-19, US and Chinese scientists partnered up to invent a vaccine for a dangerous virus that didn't yet exist. They then created such a virus in the lab to experiment on, in order to invent the vaccine. It's important to understand that that particular vaccine would only be useful if that particular virus somehow got loose. It somehow got loose.

I made a point of reserving judgment on Covid's origins until I could do some of my own extensive research and connect with knowledgeable scientists. There were a great many studies and opinions put in the public mix to muddle and confuse. But it didn't take much digging for me to unearth formidable evidence pointing to the reality.

For example, in 2020, French virologist and Nobel Prize recipient Luc Montagnier alleged that Covid's genetics were the product of "manipulation." Montagnier was cofounder of the World Foundation for AIDS Research and Prevention, and codirected the Program for International Viral Collaboration. "Someone added sequences," he said of Covid. "It's the work of professionals, of molecular biologists . . . a very meticulous work." In early 2021, Dr. Stephen Quay, MD, PhD, CEO of Atossa Therapeutics, conducted an analysis that concluded beyond a reasonable doubt that Covid-19 is lab-derived.

But the strongest evidence I found came from two sources with firsthand knowledge. They told me the US government conducted genome sequencing of Covid-19 almost immediately in the pandemic and found, among other things, clear hallmarks of man's intervention.

Funding "Gain of Function"

Beyond Covid's origins, powerful interests sought to generate a swirl of uncertainty over another pivotal question: Did US officials and taxpayer money fund the controversial research implicated? The answer: undeniably, yes. The category of research is called "gain of function." It consists of genetically engineering an organism to give it a new property. In this case, US and Chinese scientists manipulated bat coronavirus to make it infectious in people.

Gain-of-function research using potentially lethal viruses without the strictest controls is considered so risky that the US temporarily halted it in 2014. But I learned the government made a pivotal exception. The exception was a special gain-of-function study entitled "A SARS-like cluster of circulating bat coronaviruses shows potential for human emergence." It says right there in the study that the research was "reviewed and approved for continued study by [the National Institutes of Health (NIH)]" despite the gain-of-function ban.

University of North Carolina researcher Ralph Baric took a lead in this research. His biography says his work centered on manipulating coronavirus for "rapid and rational development . . . [of] . . . vaccines and therapeutics." His research partner was none other than the lead virologist at China's Wuhan Institute of Virology, Shi Zhengli, who's nicknamed "bat woman." As I mentioned, the US-Chinese team took bat coronaviruses that were harmless to humans and engineered them to turn them into genetic hybrids that were very dangerous, with a goal of inventing a vaccine that would be ridiculously profitable if the virus ever infected a bunch of people.

First, according to study documents, they succeeded in getting the nasty little viruses to infect human airway cells grafted in mice. Next, they tested a "live-attenuated virus vaccination" on the mice. "Young and aged mice were vaccinated by footpad injection . . . then boosted with the same regimen 22 [days] later and challenged 21 [days] thereafter," reported the study in *Nature Medicine* on November 9, 2015.

NIH didn't only give the research a big thumbs-up; it also paid for it with six grants of your tax dollars from the US Agency for International Development (USAID) and the National Institute of Allergy and Infectious Diseases (NIAID), led by Dr. Fauci. The grant numbers are AI085524, U19AI109761, U19AI107810, F32AI102561, K99AG049092, and DK065988. I found and published the supporting materials in June 2021. There's no doubt about any of this. There never was, despite what you may have read or heard.

So when Dr. Fauci worked to discredit the "lab theory," while failing to disclose his own role in funding the work, he was being duplicitous. It's akin to Dr. Paul Offit "debunking" safety issues with the rotavirus vaccine without mentioning he's the man who invented and made millions off the rotavirus vaccine.

Despite clear-cut documentation of US taxpayer-funded gain-of-function research with the communist Chinese, every few months media reports, news reporters, or government officials implied it was uncertain as to whether this were really true. I wondered why the skeptics didn't do a little bit of their own research or simply read the study and grants for themselves.

There was scant publicity given to another entity used to send more US taxpayer money to China for gain-of-function research. Dr. Fauci's institute and the US Agency for International Development (USAID) funneled cash through EcoHealth Alliance, a New York–based nonprofit. Ironically, EcoHealth Alliance describes its mission as "Protecting global health by preventing the outbreak of emerging diseases." Critics would argue it did the opposite.

Much about EcoHealth Alliance is shrouded in mystery. I checked out tax records and learned the group started in 1971 as a small, wildlife-centered nonprofit called "The Wildlife Preservation Trust." By 2018, the year before Covid, it had changed its name and grown into quite a different animal, collecting $18.5 million that year almost entirely from US government agencies. Its biggest single reported expense that year was about $5.43 million in grants given to foreign governments. About one-third of EcoHealth's total budget, $6.3 million, was spent on salaries and other employee-related costs. (That large proportion of money going to what's considered "administration" typically gets a nonprofit red-flagged as inefficient.) *Why were our government agencies so keen on giving to this particular nonprofit?* They already fund thousands of other global health initiatives. Why, exactly, were they providing near sole support to a group that sends millions to foreign governments?

EcoHealth Alliance's top five contributions in 2018 are listed as $11.5 million from USAID, $2.5 million from the US Department of Defense, $601,474 from the US Department of Health and Human Services (HHS), $783,412 from the US Department of Homeland Security (DHS), and $900,000 from vaccine maker Johnson & Johnson. I wondered what Johnson & Johnson's interest in EcoHealth Alliance was, and where that money went, but the company didn't reply to my multiple requests for information. A press release indicates that EcoHealth had developed an important tie with the big pharmaceutical company in 2017. It elected Marianne De Backer, vice president of Venture Investments at Johnson and Johnson, to EcoHealth's board of directors to help provide "strategic direction." Another press release in

2018 stated that Johnson & Johnson funds were being used on a project to fight deforestation in Liberia, West Africa.

At the helm of EcoHealth Alliance is Peter Daszak, a zoologist who specializes, as it so happens, in viruses transmitted from animals to people. Daszak had acknowledged in a 2015 presentation that experiments like the ones his group was involved in—infecting "humanized" mice with an engineered bat coronavirus—has the highest degree of risk. I found an online interview of Daszak with a podcast called *This Week in Virology*, recorded on December 9, 2019. In retrospect, Covid had already begun to unleash its fury on the world. We just didn't know it yet. In the online interview, Daszak discussed testing modified coronaviruses on human cells and humanized mice at the Wuhan Institute of Virology (WIV). He told the host of the podcast, "The logical progression for vaccines" is to "insert" material into the coronavirus to "get a better vaccine" and "You can manipulate [bat coronavirus] in the lab pretty easily."

Ill-Advised

As far back as 2015, independent scientists objected to America conducting gain-of-function research with China. In a published paper, a virologist at the Pasteur Institute in Paris, Simon Wain-Hobson, noted that the research had produced an engineered novel coronavirus that "grows remarkably well" in human cells and, "If the virus escaped, nobody could predict the trajectory." Biodefense expert and molecular biologist Richard Ebright of Rutgers University added, "The only impact of this work is the creation, in a lab, of a new, non-natural risk."

In 2018, US State Department science diplomats visited the Wuhan Institute of Virology (WIV). They said the lab's research on bat coronaviruses was critically important. Yet they were so concerned about safety issues that they dispatched sensitive cables to Washington, DC, warning that what they observed posed a possible risk of *starting a new respiratory virus pandemic*. In April and July 2020, the *Washington Post* published some of the contents of the two-year-old cables, which shed additional light on problems discovered earlier at the Chinese lab.

As detailed in the documents, it took China eleven years to build the Wuhan Institute of Virology. Construction was finished on January 31, 2015, and the Chinese accredited the lab in February of 2017. It houses the highest-level biosafety lab, level 4, to research "among the most virulent viruses that pose a high risk of aerosolized person to person transmission." Cables sent by US officials from the US Embassy in Beijing to Washington, DC, dated January 19, 2018, indicated, "The new lab has a serious shortage of appropriately trained technicians and investigators needed to safely operate this high-containment laboratory." According to the cables, the Wuhan lab also "has scientific collaborations" with the University of Texas Medical Branch in Galveston—once again, collaborations supported by taxpayer dollars through Dr. Fauci's National Institute of Allergy and Infectious Diseases (NIAID). Further, the cables confirmed that NIAID and the US Agency for International Development (USAID) had backed a five-year-long study on bat coronaviruses with communist Chinese scientists from the Wuhan lab, along with Daszak of EcoHealth Alliance.

After Dr. Redfield publicly leaned into the likelihood that Covid escaped from the Wuhan lab, Dr. Fauci began awkwardly spinning the news. "Obviously, there are a number of theories," he opined. "So Dr. Redfield was mentioning that he was giving an opinion as to a possibility. But again there are other alternatives, others that most people hold by." Once again, Dr. Fauci forgot to mention that he'd used our money to help pay for the very studies at issue. Instead, he steered the conversation away from anything that pointed to him or the Chinese lab. "This virus was actually circulating in China, likely in Wuhan, for a month or more before they were [*sic*] clinically recognized at the end of December of 2019," Dr. Fauci said. "If that were the case, the virus clearly could have adapted itself to a greater efficiency of transmissibility over that period of time, up to and at the time it was recognized."

I've found it's never easy to pierce secrecy surrounding government health narratives. Dr. Fauci declined my interview requests. And, as usual, the federal government failed to lawfully respond to multiple Freedom of Information Act (FOIA) requests I made for Dr. Fauci's emails, which are public documents. As I've mentioned, under FOIA law, federal agencies must hand over requested documents within about thirty days. But after waiting more than a year, I'd received nothing. Same with my FOIA requests to the National Institutes of Health (NIH). Baric, the researcher at the University of North Carolina, declined to speak with me, as did Daszak, the zoologist who leads EcoHealth Alliance. On Twitter, Daszak called the idea that Covid is connected to his research "rabbit hole conspiracies." He added, "The same gang of right wing media outlets are also posting fraudulent claims about my work. Pure politics w/out a care for how this ultimately puts public health at risk." *See? It's not Covid that put public health at risk. It's all those pesky questions about how Covid came about.*

And in case you had any doubt about how the clubby scientific community works, despite Daszak's research partnership with Wuhan lab scientists, the World Health Organization invited him to help investigate the origins of Covid (in other words: himself)! That team, with Daszak's input, issued a public report concluding it's "extremely unlikely" the virus came from a lab.

Not everybody was snookered by the subterfuge. About 1,300 people signed a petition launched by independent scientists asking the World Health Organization (WHO) and EcoHealth's Daszak to explain more about his research that was underway when the pandemic broke out. The petitioners didn't get answers.

On April 27, 2020, the Trump administration took what seemed like a logical step. It canceled the remaining funding for EcoHealth Alliance research with the Wuhan Institute of Virology (WIV). After all, the Chinese arguably cost many lives by blocking the Covid origin investigations that might have helped mitigate spread of the disease. But Trump's decision to cut off money was met by a political backlash. Establishment scientists and their friends in the media insisted the move was irresponsible. Pretty soon, the National Institutes of Health (NIH) reinstated the EcoHealth grant. But there were a few strings attached. For EcoHealth Alliance to get the cash, it would have to convince its Chinese partners to finally answer questions about the lab's practices and the Covid outbreak, and provide a virus sample. *Foul!*

cried EcoHealth Alliance and its allies. They claimed the conditions made research "impossible" because the Chinese would never agree.

Unbelievably, on August 27, 2020, it was announced that NIH had awarded an even larger grant of taxpayer money than before to EcoHealth Alliance: $7.5 million. EcoHealth Alliance was one of eleven institutions and research teams approved to receive an $82 million bundle of US tax money to study viruses crossing from nature into people and develop rapid-response strategies. Talk about throwing good money after bad! I cannot say with any certainty that this commitment of money amounts to a payoff to keep EcoHealth Alliance quiet about what it knows. But it's reasonable to wonder. *Congratulations! You failed so hard at preventing or mitigating the Covid pandemic, we're rewarding you with even more money.* In May 2024, after an embarrassing performance by Daszak two weeks earlier at a congressional hearing, the government finally decided to suspend federal grants to EcoHealth Alliance. Both Republicans and Democrats criticized EcoHealth for failing or delaying to properly report its high-risk studies.

True "Scientific Consensus"

If you take a little time and get beyond Google, Wikipedia, the CDC, the World Health Organization (WHO), public health officials, and establishment media, you'll find there were many people unearthing the truth about Covid's origins and gain-of-function research. Often, propagandists tried to claim that any authorities pointing to the Wuhan lab were "right-wing," or Trump supporters. But many were not.

Take Jamie Metzl. He's a member of the World Health Organization (WHO) International Advisory Committee on Human Genome Editing. He served as deputy staff director of the Foreign Relations Committee under Senator Joe Biden on the National Security Council, and at the State Department under President Bill Clinton. In 2021, he told me that his own research into Covid's origins had turned up many of the same answers as my research had revealed.

As part of my research, I consulted numerous scientific sources who'd established credibility with me in the recent past by proving to be accurate on key points of scientific debate. On the issue of sensitive US gain-of-function research with China, each one told me it should never have been allowed. One of the sources, a medical doctor, said it was "irresponsible" to "partner with China on how to make [coronavirus] more infectious." Another, also a medical doctor and biodefense expert, said, "Hell, no, it's not a good idea . . . [China has] an active bioweapons program, a very good one . . . and you're going to cooperate with them on gain-of-function research? Somebody's IQ dropped sharply when that decision was made."

"There are scientists all around the world who have told me that they believe the most likely origin of Covid, of the pandemic, is an accidental lab leak from the Wuhan Institute of Virology," Metzl tells me, yet another expert voice contradicting Dr. Fauci.

"What have you been told, and what have you found about scientists who feel like they can't step forward?" I ask.

"Many of these people are afraid to step forward," Metzl replies. "They've called it 'career suicide,' because there are so many contentious issues, because the stakes are so high. Because the Chinese government, in collaboration . . . with some very high-level and prominent scientists have put forward this story that I think is wrong."

Metzl says Daszak at EcoHealth Alliance drove efforts to controversialize the mere asking about whether Covid could have come from a lab. And Daszak was part of the campaign to label reporting about it as "crackpot theories."

"I have repeatedly called for Peter Daszak to be removed from the WHO Organized International Advisory Committee looking into the origins of the pandemic," Metzl tells me. "And the reason why I have done so is Peter has a tremendous conflict of interest as someone who is, through his organization, the EcoHealth Alliance, a significant funder of gain-of-function research at the Wuhan Institute of Virology."

As part of his investigations, Metzl uncovered a familiar tactic put to use to deflect scrutiny from China and its US supporters. The unseen hands of Daszak and his colleagues used medical journals to discredit the "lab theory." They worked vigorously behind the scenes to orchestrate a letter published February 19, 2020, in the prestigious

British journal *Lancet* entitled: "Statement in support of the scientists, public health professionals, and medical professionals of China combatting COVID-19." Documents later obtained though Freedom of Information requests show that while Daszak was enlisting people to sign on to the *Lancet* statement, he sent emails urging that the statement should "not be identifiable as coming from any one organization or person" but seen as "simply a letter from leading scientists." As the science journalist Thacker later observed in *The Disinformation Chronicle,* Daszak was "moving behind the scenes to create this public outcry in the *Lancet* that anyone looking into a lab leak is a conspiracy theorist."

The final *Lancet* letter read in part: "We stand together to strongly condemn conspiracy theories suggesting that COVID-19 does not have a natural origin." Daszak and the other signatories forgot to mention their own ties to the Chinese lab in question. And their letter was widely quoted in the press as evidence that Covid-19 had a natural origin. In a follow up article in *The Guardian,* Daszak calls the idea of lab origin for Covid a "crackpot theory." *Settled Science.*

"This letter [in *Lancet*] was considered, at the time, very credible," Metzl notes. "There were a number of Nobel laureates who signed it. And only later did it come out through a Freedom of Information request that the entire process had been managed and manipulated."

An interesting postscript. Jacques van Helden is a professor of bioinformatics at Aix-Marseille Université (AMU) in Marseille, France, where he specializes in analyzing genomes and genome regulation, and is co-director of the Institut Français de Bioinformatique, which supports research in life sciences. Van Helden told Thacker that he contacted *Lancet* to urge them to publish his own letter refuting the Daszak-organized letter.

"I was told [by *Lancet*] our letter would be evaluated by a special committee on the virus origins," he says. "This special committee has a president, and it is Peter Daszak." Van Helden says he told the *Lancet* editor, "OK, but do you realize there may be a conflict of interest? He is the person who wrote the statement in the *Lancet* that we are saying was wrong . . . [The editor] didn't answer that . . . they decided to uphold their position that [our] letter would not be considered [for publication]."

Let that sink in. The head of the special committee advising on Covid's origin was a man whose research was implicated in Covid's origin.

For her part, Shi, the "bat woman" at the Wuhan Institute of Virology, has been firm in her denial of anything to do with Covid-19, calling the virus "nature's punishment on the human race." "I swear on my own life that the virus has no connection with the laboratory," Shi insisted in a statement. "To those people who believe in and are spreading the rumours perpetrated by third-rate media outlets . . . I would like to give this advice: Shut your dirty mouths!"

While some of us were busy following Covid's trail, there appeared to be an organized effort to undercut any studies that supported the lab theory. Acting apparently in tandem, medical journals added an identical "editors' note" to the inconvenient studies. *Don't believe what you're about to read in this study*, the editors' note implies. *This study doesn't say what you think it says*. The note, still appended to the studies today, reads: "We are aware that this article is being used as the basis for unverified theories that the novel coronavirus causing COVID-19 was engineered. There is no evidence that this is true; scientists believe that an animal is the most likely source of the coronavirus."

More Manipulation

Email evidence released in 2022 and 2023 shows that Dr. Fauci, the head of the National Institutes of Health, Dr. Francis Collins, and the bat coronavirus researcher at the University of North Carolina, Ralph Baric, also worked together to use medical journals to spin the public about Covid's origins. They secretly weighed in on or helped direct publication of articles that pointed to a natural origin for Covid and resulted in deflecting from Covid research they funded or took part in with the Wuhan lab. Investigative science journalist Thacker, who reports on corruption in science and medicine, spoke with me about some of the emails. He says Drs. Fauci and Collins violated their own established ethics rules with their furtive acts.

By way of background, Thacker gives me an interesting bit of trivia about the beginnings of Fauci's institute. "The thing you have to understand is that while it's called the 'National Institute of Allergy and Infectious Diseases,' it's actually a biodefense research institute housed within the National Institutes of Health," says Thacker. "That's not spoken about much today. . . . So what he really runs is a biodefense institute. . . . I think that if Anthony Fauci had been brought on television or in White House [Covid] briefings and had been identified as 'head of biodefense research' and not as 'head of an Institute of Allergy and Infectious Diseases,' I think this would have created a different perception with the American public."

So if we want to examine the existence of Covid from a biodefense perspective, the question becomes: *How does one create a defensive weapon, such as a vaccine, against a biological agent that doesn't yet exist?* Answer: *By creating that agent.* As I've explained, that's exactly what the US government and China did together with the gain-of-function research. "And so that creates this whole line of research in which we're taking these viruses and ginning them up and making them more dangerous so that we can then create defensive countermeasures to them," says Thacker.

As far as the US government's explanation for why they were conducting the ill-advised research, it never made sense to me. At the 30,000-foot level view, it defies logic. How does it make any sense to invent a dangerous virus in the lab that may never come into existence on its own, in order to be prepared for it? What are the odds that even the best scientists in the world could accurately predict what would be the next virus that nature would unleash upon mankind, among the infinite possibilities? I'm not sure we'll ever know the full story behind what was really going on.

In any event, when Covid broke out, most likely from the very Chinese lab the US had collaborated with and funded, the responsible parties went to great lengths to divert blame.

"We have a sequence for the virus that's released January 11, 2020," says Thacker. "We then see a series of [scientific] papers that come out that say it couldn't have possibly come from a lab." One was written and published in the journal *Emerging Microbes and Infections.* Emails later showed that the authors shared an advance draft of the article with the scientist leading the gain-of-function research in question: Baric.

"[Baric] edited their piece and then told them explicitly in an email, 'I don't want my name on this,'" says Thacker. An advance draft of the article was also shared with bat woman Shi. And her comments were incorporated. "They then published this piece without disclosing that there had been this secret editing by Ralph Baric and by Shi Zhengli," says Thacker.

But wait! There's more. There was also a letter called "Proximal Origin," authored by five scientists and published in *Nature* in March 2020. It too claimed that Covid-19 couldn't have come from a lab. And this letter bore the fingerprints of Drs. Fauci and Collins.

"What we know now is that the people who wrote [the "Proximal Origin" letter in *Nature*] initially were very concerned that [Covid-19] may have been engineered," says Thacker. But then there was a fateful behind-the-scenes teleconference. Emails indicate the teleconference included Drs. Fauci and Collins, as well as Jeremy Farrar, who ran the Wellcome Trust–funded Oxford University Clinical Research Unit, a giant funder of biomedicine. After the teleconference, the authors of "Proximal Origins" began sharing multiple drafts with Fauci, Collins, and Farrar. The lead author, Kristian Andersen, thanks "Jeremy, Tony, and Francis" for their "advice and leadership" and ultimately sends them the accepted version of the manuscript, as well as a draft press release. Andersen asks the shadow advisors if they have "comments, suggestions, or questions about the paper or the press release."

Thacker says that after input from Drs. Fauci and Collins, the article got a miraculous conversion. It said Covid-19 is natural and couldn't have come from a lab. And after all that backdoor "advice and leadership," the role of Fauci and Collins wasn't even mentioned in the article. "After Francis Collins and Tony Fauci helped to orchestrate this piece and plant it in *Nature*, they then amplified it themselves," says Thacker. "Francis Collins a couple weeks later writes about it . . . for his NIH director's blog . . . no disclosure of . . . his advice and leadership on it. And Anthony Fauci discussed it in a White House briefing where President Trump was asked about 'How did this pandemic begin?'

Tony Fauci steps up and says, 'Oh well, we have this paper by these, you know, international virologists.' No mention at all of his work behind the scenes to help to orchestrate that paper."

By the way, Farrar, who was part of those consultations to deflect from Covid's lab origins, got a new job in 2023. He was hired as chief scientist at the World Health Organization.

Twisted Tale

Another Covid-related science scandal that Paul Thacker uncovered in 2021 exemplifies how upended science and news reporting have become, and how those who practice true journalism can find themselves targeted by the censorship-industrial complex. Thacker published a groundbreaking investigation. Its title alone, "Pfizer's Problematic Vaccine," was enough to send a chill down the spine of news outlets drinking liberally from the spigot of pharmaceutical dollars. There was a time when journalism like Thacker's would have been published in the *New York Times* or covered on *60 Minutes*. But no more. Instead, Thacker only found a home for his reporting at the *British Medical Journal* (*BMJ*).

Thacker's investigation centered on a whistleblower at a contractor running some of Pfizer's clinical vaccine studies, Ventavia. Thacker traced what happened after that whistleblower, Ventavia regional director Brook Jackson, reported major lapses in Pfizer's Covid-19 vaccine clinical trial to the FDA. There were serious issues with "poor laboratory management, patient safety concerns, and data integrity issues." The company fired Jackson the same day she made the report.

Thacker's article about this was meticulously documented and sourced. He interviewed Jackson on the record—a credible, high-ranking official providing verifiable, firsthand information. Jackson, in turn, provided *BMJ* with dozens of internal company documents, photos, audio recordings, and emails. They show she repeatedly notified her company of the failings she discovered, and that after she reported them to the FDA, the agency never investigated. One email was about a scheduled "Clean up Call" to fix data on Pfizer's study. "[She] also had secret recordings," says Thacker. "Jackson's company started lying to reporters, saying she was never on the Pfizer clinical trial. But we had internal emails proving she was."

Thacker followed up the reporting with separate articles in his Substack newsletter, "The Disinformation Chronicle." He highlighted the enormous profits Pfizer made from its Covid vaccine after the questioned studies. The story generated news reports in the US, Australia, Italy, the UK, and South Korea. However, powerful interests worked to discredit the whistleblower, dismiss Thacker's reporting, and censor the article from the public's view.

As part of that campaign, Facebook's supposed "fact-checker," "Lead Stories," stepped in and began discrediting and censoring Thacker's article. To me, Lead Stories seems to further government and pharmaceutical industry propaganda rather than check facts. It slapped a label on Thacker's article warning: "Missing context: Independent fact-checkers say this information could mislead people." Facebook also added instructions to posts that referenced the article urging, "Visit the COVID-19 Information Center for vaccine resources." That linked users to Facebook's propaganda-laden Covid center. To its credit, *BMJ*, which published Thacker's investigation, responded by sending Facebook's Mark Zuckerberg an open letter. *BMJ* called Facebook's "fact check" "inaccurate, incompetent and irresponsible." A cry in the dark, perhaps, but at least they didn't take it lying down.

In 2023, some telling emails were released. They revealed some of the hidden machinations that were behind Thacker's article getting flagged on social media. The emails were turned over in a lawsuit filed by Missouri and Louisiana against the Biden administration for Covid censorship.

"[M]onths before *BMJ* published my investigation, Facebook cut a deal with the Biden White House to limit 'true content' on vaccines," Thacker notes. He's referring to an email exchange between Facebook and White House officials Andrew Slavitt and Rob Flaherty dated March 21, 2021. In the emails, an unnamed Facebook official assures the White House duo that Facebook is firmly on Team Vaccine even when it means hiding the truth. Facebook tells the White House: "[W]e have been focused on reducing the virality of content discouraging vaccines that does not contain actionable misinformation. *This is often true content* [emphasis added] . . . we'll remove these Groups, Pages, and Accounts. . . ."

Those revelations about White House involvement in social media censorship should have spurred a national movement to address the perfidious collusion between government, industry, and information brokers. But therein lies the dilemma: the same players who control the information . . . control the information about their control of the information.

In 2022, my co-judge, Professor Alberto Martinez of the University of Texas, and I chose Thacker's reporting on Pfizer for first place in the cash prize journalism awards program I'd just started, The Sharyl Attkisson ION Awards for Investigative and Original News. (If you bought this book, a portion of your money is going toward the awards to encourage independent, accurate reporting.) The idea behind why I created the award is that some of the most deserving journalism today will never be recognized by the well-known establishment groups, for a variety of reasons. People serving as award judges for those groups tend to be other journalists who are loath to give positive recognition to off-narrative reporting, particularly that which questions their own narratives. Instead, they frequently reward slanted reporting, even when it proves to be false and irresponsible. An example can be found in the *New York Times*' unforgivably sloppy, error-ridden, and biased reporting that claimed President Trump was a Russian stooge. It won a Pulitzer Prize. But news stories that accurately reported on the fraud behind the *Russia, Russia, Russia!* claims are considered "discredited." Additionally, many legacy journalism awards groups—like every form of influence and information that *can* be co-opted—*have* been co-opted, by the usual suspects.

Meantime, in 2021, a group called "The Virality Project" was joining the Biden administration in advising Twitter to follow the same path as Facebook in censoring even perfectly "true posts which could fuel [vaccine] hesitancy."

The Virality Project

The case of the Virality Project provides an additional window into the insidious nature of information manipulation in the name of science. We were offered a glimpse into this propaganda effort in 2023, as a result of Elon Musk's takeover of Twitter. Musk released the so-called Twitter Files, which included communications between pre-Musk Twitter and disinformationists at the Virality Project.

The Virality Project was started around January 2021. It described itself as "a coalition of research entities focused on real-time detection, analysis, and response to COVID-19 anti-vaccine mis- and disinformation." But the group's work product reads much more like a vaccine-allied effort to suppress off-narrative facts and opinions. The group worked to quash admittedly true information about Covid vaccine side effects, breakthrough infections in vaccinated people, and the power of natural immunity. It pressured Twitter to squelch criticism about vaccine passports, mandates, lockdowns, and masking. It tried to eliminate news about other countries restricting Covid vaccines, warning about their safety, or questioning their effectiveness. And it was responsible for furthering a whole lot of misinformation. For example, Virality Project analysts labeled as "misinformation" claims that vaccines didn't prevent Covid from being transmitted—after even the CDC admitted that was true.

Most "incidents" tagged as misinformation by the Virality Project were nothing more than prominent voices who were off the government script. For example, "Virality Project Weekly Briefing #26," dated June 16 through June 22, 2021, devoted ink to the personal choice of NFL player Cole Beasley to skip the Covid vaccine. The Virality Project quantified how many "engagements" Beasley's statement received on Twitter ("over 115K") and noted with concern that Beasley's language aligned with "both the medical freedom community and the 'natural immunity' narratives."

"Takeaway," concluded the Virality Project analysis: "Prominent athletes have the potential to influence public opinion, and [Beasley's] perception of a lack of risk to himself as a healthy man may resonate with his fan base." That's something the Virality Project could not permit: a free thinker drawing his own conclusions.

Propagandists working on the Virality Project were long on censorship but short on accuracy. In another briefing, they proclaimed, "This week we saw the censorship discussion take a partisan turn. Breitbart News and Fox News host Tucker Carlson, prominent anti-vaxxer Robert F. Kennedy Jr., and foreign media outlets [*sic*] RT pushed similar stories about the White House's relationship with social media." Yet in truth, there was little partisan agreement at the time among the groups named—quite the opposite. Breitbart News typically comes from the right, while Tucker Carlson could be said to lean anti-establishment and libertarian-conservative, and Robert F. Kennedy Jr. was a lifelong liberal Democrat. When such a diverse group began to express similar criticism of White House censorship efforts, it threatened to sway the general public. The Virality Project's effort to brush off agreement among them as "partisan" was itself an exercise in disinformation.

I've had enough experience with those hired to spin and smear that I have come to recognize their handiwork. The Virology Project used the expected, hallmark buzzphrases in describing its work. It claimed it "supports information exchange between public health officials, government, and social media platforms through weekly briefings and real-time incident response." *Information exchange. Incident response. Code words also used in vaccine propaganda.* Their real mission appeared to be to connect, pressure, and influence government, public health officials, and social media. It's fair to conclude that, in their view, certain truths must never be told, lest the public form conclusions undesirable to The Goal.

In peeling back the onion layers, we discover that your tax money is helping finance the censorship. For example, the Virology Project was led by researchers at Stanford University in partnership with several nonprofits that received government funds. Further, as Reason.com wrote, "Some of the Virality Project's partner organizations received funding from the Department of Defense and the National Science Foundation. The project also said it had developed 'strong ties' with federal government agencies." The arrangement illustrates an uncomfortably common closeness among government, media, and corporate interests.

When I poked around, I also found some familiar, nongovernment names involved in the Virality Project. One "sponsor" is the University of Washington "Center for an Informed Public," which could be more aptly named "Center for a Propagandized Public." It's funded by the same groups that frequently support one side of social and political controversies and work to discredit the other. One such funder is the Knight Foundation, a prolific supporter of vaccine propaganda that labels legitimate concerns as "anti-vaccination"; woke efforts such as those involving transgender causes; and controversial climate change initiatives. Knight Foundation is also connected to the Soros Open Society Foundations, one of the largest backers of global woke causes and information manipulation. Another example of a powerhouse behind the scenes at the "Center for an Informed Public" (and, by proxy, the Virality Project) is the Hewlett Foundation, started by Hewlett-Packard cofounder William Hewlett and his wife. It sponsors left-wing causes such as workshops and research projects at the University of California, Berkeley, on climate change, inequality, and "the relationship between capitalism and democracy." All of the aforementioned took controversial positions heavily promoting the experimental Covid vaccines and made no easily identifiable effort to balance the advice with reasonable cautionary notes. All of these influences are hovering somewhere in the background of the Virality Project.

The Virality Project propagandists also supported a controversial new government censorship agency established under the Department of Homeland Security during this time frame. Its stated purpose was to root out "misinformation, malinformation and disinformation." The name, the "Disinformation Governance Board," was eerily evocative of the "Ministry of Truth" in George Orwell's novel *1984*. The Ministry of Truth, you may recall, was all about lies. It was the place where workers falsified records to force history into agreement with the ever-evolving positions of all-powerful "Big Brother."

As if the mere idea of a government-run Disinformation Governance Board weren't bad enough, Americans weren't told about its creation until after it had already been quietly operating. In early 2022, the public was finally clued in. Homeland Security Secretary Alejandro Mayorkas announced a controversial choice to head it up: Nina Jankowicz.

Jankowicz added another facet to the scandal. She was an odd choice for executive director of the Disinformation Governance Board if truth were the goal, critics said, because she was well-known for dispersing disinformation. An advisor to Ukraine, she'd reportedly joined with voices who falsely labeled Hunter Biden's laptop "Russian disinformation." It didn't help her standing when she posted a bizarre TikTok social media video singing parody lyrics to the tune of "Supercalifragilisticexpialidocious" from the film *Mary Poppins*. In the embarrassing performance, an overexcited, hyperventilating Jankowicz sang out attacks against Trump supporter Rudy Giuliani, and predictably implied that people who were off the government Covid narrative could find themselves in her crosshairs. After a bipartisan outcry, Jankowicz resigned from her post and Mayorkas disbanded the short-lived Disinformation Governance Board. Many of us have little doubt that the government is still policing free speech through other means and methods.

We've touched upon only the briefest sliver of the giant propaganda and money machine devoted to controlling our information about Covid, science, and beyond. The fact that the machine has proven so effective in a country where free speech is written into the Constitution should be of concern to all of us. It's most definitely a call to action.

CHAPTER 13

The Amish Approach

When it comes to scientific inquiry conducted to advance human health, if it isn't likely to result in a boon for the pharmaceutical industry, the scientific establishment may go to great lengths to block it. A salient case in point is the Amish approach to Covid.

The Amish are a Christian group that emphasizes the virtuous over the superficial. They don't usually drive, and don't routinely use electricity or have TVs. And during the Covid-19 outbreak, they became subjects in a massive social and medical experiment.

"There's three things the Amish don't like," Calvin Lapp tells me on my mid-2021 visit to the biggest Amish community in the United States: Lancaster, Pennsylvania. Lapp is Amish Mennonite. "And that's government—they won't get involved in the government. They don't like the public education system—they won't send their children to education. And they also don't like the health system. They rip us off. Those are three things that we feel like we're fighting against all the time. Well, those three things are all part of what Covid is."

Lapp is my point of access into the uniquely Amish response to the Covid pandemic. After a brief shutdown in the beginning, the Amish chose a different path that led to Covid tearing through the community at warp speed. It began with an important religious holiday in May 2020.

"When they take communion, they dump their wine into a cup,

and they take turns to drink out of that cup," Lapp explains. "So, you go the whole way down the line, and everybody drinks out of that cup. If one person has coronavirus, the rest of the church is going to get coronavirus. The first time they went back to church, everybody got coronavirus."

Lapp says they weren't denying Covid. They were facing it head-on. "It's a worse thing to quit working than dying. Working is more important than dying," he says. "But to shut down and say that we can't go to church, we can't get together with family, we can't see our old people in the hospital, we got to quit working? It's going completely against everything that we believe. You're changing our culture completely to try to act like they wanted us to act the last year, and we're not going to do it."

That also meant avoiding hospitals. The Amish simply refused to go to the hospital, even when they were very sick, because that would mean they couldn't see visitors. And, to them, they'd rather be very sick at home with people around them than be isolated in a hospital.

The Amish anecdotes were powerful, but solid proof wasn't easy to find because the Amish didn't typically take Covid tests. Their thinking, says one observer, was, "I'm sick. I know I'm sick. I don't have to have someone else telling me I'm sick."

"We didn't want the [Covid positive test] numbers to go up, because then they would shut things more. What's the advantage of getting a test?" explains Lapp. They also didn't mask. Or vaccinate. "Oh, we're glad all the English people got their Covid vaccines," he continues, referring to non-Amish as "English." "That's great. Because now we don't have to wear a mask, we can do what we want. So good for you! Thank you. We appreciate it. Us? No, we're not getting vaccines. Of course not. We all got the Covid, so why would you get a vaccine?"

Then, in March 2021, came remarkable news. The Associated Press and other media outlets reported on claims that the Lancaster County Amish were the first to achieve "herd immunity." That would mean that a large part of the population had been infected with Covid-19 and become largely immune to it. Here was ready-made data that would allow scientists to compare what happened in this population that took no special steps to fight Covid, to those that shut down tightly and used respirators, vaccines, masks, and more. But, oddly, no establishment scientists seemed to want to fairly examine the data.

When I dug in to investigate the Amish approach to Covid, one thing became clear. Whether looking at anecdotes or coroner numbers, there was no evidence of any more deaths among the Amish than in places that shut down tightly. Some claim there were fewer. And instead of obliterating their economy the way most of the world did with mandatory shutdowns and pressure to isolate, Lapp says the Amish stayed completely open, and made more money as a community than ever before!

I aired my report about the Amish on my Sunday television program *Full Measure* in October 2021. It was of huge viewer interest at a time when the public was feeling the heavy hand of censorship, and was thirsty for accurate information about natural immunity and alternatives to the shutdowns. On my program's unadvertised YouTube channel, a replay of my TV story on the Amish quickly racked up 100,000 views. Then a million. Then three million. In a normal media environment, such a report would be picked up by news sites around the world. In a normal scientific environment, the full weight of the research community would put its efforts into learning more about the Amish approach, and launching studies to find verifiable data. But that's not what happened.

Lapp told me he'd braced himself, after my report, to field calls from the *New York Times* and the news networks wanting to follow up on the story. Nobody, not even one media outlet, contacted him.

What did the establishment research community do? They set about to try to throw cold water on the Amish approach, without actually collecting any real on-site data or conducting interviews. A year later, when I wanted to update my story, I contacted a Mennonite history professor who had provided comment for my original report. He had verified the Amish rejection of CDC-recommended practices. But his response to me now, a year later when I followed up, made me wonder whether he'd fallen under pressure after being in the first episode, which went viral. In an email to my producer, the professor now declared that "there was no herd immunity protection" among the Amish. Of course, he was wrong. Had he been correct, that a large group of people infected by a virus didn't develop immunity, it would be a history-making scientific anomaly that would force a rewrite of everything virologists believe about how viruses and acquired immunity work! In fact, most scientists and published studies were already showing that Covid natural immunity was providing the predictable superior and longer-lasting immunity than the vaccines. Even the CDC, bringing up the rear, acknowledged that people infected with Covid appeared to have "some protection" against repeat infection and/or serious illness. Scientific principles have long held that "herd immunity," by definition, is achieved faster in a population that spreads a virus faster; slower in a population that doesn't. Quicker herd immunity is the logical outcome for any population that were to follow the Amish approach.

Next, the Mennonite professor argued in his email that the Amish approach to Covid had failed because he said "Amish excess deaths nationally shot up . . . from September to November of 2021 . . . matching the national pattern in deaths." He seemed to have no idea that he was making the opposite point than what he intended. If Amish death rates truly "matched the national pattern" while the Amish avoided shutdowns, masking, isolation, experimental vaccines, and all the expense—then wasn't the Amish approach proving superior? Second, if the Amish deaths truly "shot up" for that short period, equalling the national pattern—doesn't that mean their deaths had been lower than the rest of the nation for a critical time prior to that? And third, it's unclear what data the professor was using to make his claim about the number of Amish Covid deaths since nobody tracked them with any precision.

Besides the Mennonite history professor, there seemed to be others who were determined to incorporate a revisionist narrative into the fabric of the Amish Covid story. Their handicraft came in the form of a study published in the *Journal of Religion and Health* on June 11, 2021. It was titled: "Closed but Not Protected: Excess Deaths Among the Amish and Mennonites During the COVID-19 Pandemic." In a convolutedly worded conclusion the authors wrote, "The excess death rate for Amish/Mennonites spiked with a 125% increase in November 2020. The impact of COVID-19 on this closed religious community highlights the need to consider religion to stop the spread of COVID-19." They clearly intended to leave the impression that the Amish and Mennonite suffered a far worse fate for having rejected CDC recommendations. But that's untrue. And it didn't take a lot of digging to find fatal flaws in the study.

Keep in mind that when research is launched in order to prove a desired outcome, as it frequently is today, you can assume the scientists make choices along the way to drive the results. It's hardly scientific, but, as I've described, that's just the way it is. Sometimes, bias comes in the form of reverse engineering. By that, I mean that to get to a particular conclusion, the researchers toy with a variety of datasets, populations, locations, and dates until they find a scenario that comes closest to proving their desired point. Then they use that combination of datasets, populations, locations, and dates in their "study."

In "Closed but Not Protected," some of the processes researchers followed to get to their conclusions were unclear to me. It certainly raised questions in my mind as to whether they were operating with the sort of bias that commonly plague today's science. When I contacted the lead author, she refused to answer some important questions. The first obvious problem I saw is that they failed to study the very population at issue—the Amish. Instead, they studied a confusing conglomeration of Amish and Mennonite. *How did this problem escape the study reviewers?* It matters because Mennonites are more likely than Amish to live lives closer to those of ordinary Americans and follow public health recommendations. For example, almost all the residents living at two Mennonite Home Communities in Lancaster, Pennsylvania, got vaccinated for Covid. According to the facilities' websites: "Numerous vaccine clinics have been held at [Mennonite Home] and [Woodcrest Villa], resulting in almost all residents being vaccinated against COVID-19 infection." These Mennonite residents also wore masks, isolated, and followed CDC protocols, including social distancing. So, in "Closed but Not Protected," the scientists may as well have been analyzing the regular CDC-compliant population, not the uniquely Amish approach. Their conclusions didn't necessarily reveal anything about the Amish approach.

A second issue with "Closed but Not Protected" can be found in the geographical choices the researchers made. Though their conclusions made sweeping generalizations about the Amish approach, they had, in fact, omitted the most populous Amish community in America: Pennsylvania. Instead, they focused their attention exclusively on "Ohio, with the second and fourth largest Amish settlements in the USA."

When I asked study author Rachel Stein at West Virginia University why she excluded Pennsylvania, she provided a surprising answer. She indicated she and her colleagues chose Ohio because it had a disproportionately high number of Amish and Mennonite obituaries compared to Pennsylvania and other states. According to Stein, Ohio is "home to approximately 23% of the US Amish population, but contributing 56% of the total obituaries published" and "Pennsylvania was not represented to the same degree as Ohio in the data." Incredibly, this seems to mean that when they saw evidence of a more positive outcome in Pennsylvania—fewer obituaries—they excluded that data from their study. Why did the researchers ignore the obvious possibility: that the Pennsylvania Amish and Mennonite had fewer published obituaries compared to Ohio *because their death rate was lower*? In effect, the decision to omit Pennsylvania would seem to artificially elevate the apparent death rate among Amish and Mennonite people and present the worst case outcome for Closed Religious Communities (CRCs).

A third flaw with the study is the strange mismatch between the grant proposal, submitted in order to obtain the $258,000 government grant, and the study itself. They seem to describe two entirely different projects. The proposal says researchers intended to examine "the prevalence of COVID-19 related health information and misinformation and social distancing and isolation practices within closed religious communities" and "whether group closure and cohesive organizational network ties are associated with an increased prevalence of health misinformation and a reduced prevalence of social distancing and isolation practices." Yet the study didn't seem to focus on any of that. Nor did it define what it considered to be "health misinformation." That's a critical question, since the CDC itself distributed so much misinformation. We need to know what the study authors considered to be misinformation. Further, the proposal stated that researchers wanted to compare the Amish to the Mennonite population, with the full understanding that the Mennonite are more likely to follow CDC-recommended health advice. Why did they end up blending the two disparate groups together instead of comparing them? And finally, the grant proposal said researchers intended to analyze "scribe reports by congregational members . . . for COVID-19 related information and social distancing and isolation content" and track "COVID-19-related official instructions . . . over time." But there was no evidence the researchers had studied any of that, either. When I raised the issues with Stein, she said the outstanding questions would be addressed in other related studies.

I think the most significant flaw with the study is that it buries an earthshattering finding—one that's contrary to establishment science narratives: Even using slanted data that likely exaggerated the toll that Covid took on the Amish, the researchers found no evidence the Amish suffered any worse than the rest of America. At the same time, the Amish managed to avoid shutdowns, isolation, masking, testing, hospitalization, and vaccination.

The Amish approach appears to be far superior.

Following a Murky Trail

Since "Closed but Not Protected" was funded by our taxpayer money through the National Science Foundation, details about it belong to us. We paid for it. We own it. So it should be easy for us to find answers to key questions about the researchers' approach and who may have guided it. But when I raised issues with the publishing journal, the *Journal of Religion and Health*, editors sent me in a loop, referring me back to study author Rachel Stein, who had gone dark when I persisted with queries.

I wondered why a legitimate researcher would seek to keep aspects of her published work hidden from public view? In this instance, I thought it would be instructive to find the names of the "peer reviewers" who approved the study for publication. Could there be conflicts of interest among them? Had players in government, or other known propagandists in the scientific community, had a hand in green-lighting a seriously flawed study to further a pro-vaccine narrative?

Apparently, we'll never know. When I filed a Freedom of Information Act (FOIA) request for the names of the reviewers, the National Science Foundation (NSF) informed me of the following: "The NSF does not release reviewer names; accordingly, redactions will be made to protect the personal privacy of the reviewers under Exemption (b)(6) of the FOIA." In other words, the government has granted itself the right to spend our money on a study, then block us from being able to conduct reasonable oversight of the money they spend.

West Virginia University eventually withheld all the relevant communications about the study and emailed me a set of documents that mostly consisted of my own emails sent to Stein. The only bit of insight provided among them were a few emails from other observers who had written Stein with similar questions about facets of her study.

I set forth my concerns about the Amish study in a letter to the *Journal of Religion and Health* since corrections and a retraction seem to be in order. The *Journal* assured me the concerns would be investigated. Eleven months later, no response had been provided.

CHAPTER 14

Teaching to the Choir

Medicine on the Take

Regular Americans have no idea of just how much the education of our doctors is dictated by the pharmaceutical industry. A case in point is the Continuing Medical Education classes that medical professionals take to maintain their licenses and practices. One would expect the courses to be strictly separated from industry influences. But at this point in the book, you can guess the reality.

On May 31, 2023, the industry website Medscape, which has a left-leaning, pro-pharma tilt, posted Continuing Medical Education materials for a class about Covid in children. *I'm interested.* I'd begun to wonder what doctors are being officially taught now that so much evidence has emerged confirming the resilience of kids who get Covid infections and elaborating on Covid vaccine risks. Are doctors being told the fact that many experts believe the risks of getting vaccinated outweigh the supposed benefits for most children?

The same week the Continuing Medical Education class materials are posted, there happens to be reassuring data from Israel about Covid in young people. Israeli health officials were unable to point to a single Covid death among otherwise healthy people under age fifty. Though there was a limited dataset, the findings serve in strong support of what's becoming widely accepted: Covid isn't a serious disease among otherwise healthy people. *How dare you say that?* retorts a small army of propagandists assigned to adjust the narrative. *Healthy, young people* do *die of Covid! Don't listen to dangerous* "anti-vaccine grifters"*!*

As I review the slideshow for the Continuing Medical Education class, I quickly recognize the same vaccine industry slant that's become so familiar to me in more than two decades of investigative reporting on the subject. The materials contain no information on what's probably the most important question: What's the risk from getting Covid compared to the risk from getting the vaccines? There's also no admission that scientists won't know for decades the true impact and range of vaccine side effects. There's no examination of the most serious vaccine outcomes, including paralysis and death. And there's no acknowledgment of important children like poor Maddie de Garay of Cincinnati, Ohio.

Maddie's Story

It was summer of 2020 when a school friend told Maddie de Garay's brother about an exciting opportunity. A clinical trial was about to be launched at nearby Cincinnati Children's Hospital. It would test Pfizer's experimental, hastily developed Covid vaccine on twelve- to fifteen-year-old children. Young volunteers were needed. The kids who volunteered would be paid a hundred dollars or so per visit.

The de Garays didn't hesitate to consent to all three of their children taking part. They were excited to contribute to the greater good, and glad to have their kids be among the first to get a lifesaving vaccine—or so they thought. Their mom, Stephanie, later told me she didn't think twice. After all, she thought, everyone knows vaccines are safe. Besides, she figured, any kids in the clinical study would get the closest scrutiny and care.

Among the three siblings, it was twelve-year-old Maddie alone who happened to be chosen to get two doses of Pfizer's vaccine, rather than a harmless placebo or a combination of vaccine and placebo. This proved to be a fateful and debilitating roll of the dice. Within hours of her second dose on January 20, 2021, Maddie fell apart. That night, she woke her parents and climbed into bed with them, saying she didn't feel right. She went to school the next day but barely made it home on two feet, complaining about feeling electric shocks going up and down her spine.

"She said she felt like her heart was being ripped out," Stephanie tells me when I visit the family's home in 2023. "Chest pain. She had severe abdominal pain. She was hunched over when she walked through the door. Her toes and her fingers were white, and when you touched them, they were ice-cold and painful." That began a long nightmare for the whole family. Stephanie says she'd assumed if any of her kids had health issues during the vaccine research, Pfizer and the doctor leading the study would pull out all the stops to help. But when Maddie got sick, she says the opposite happened. To her, the Pfizer folks appeared only interested in blaming their once-healthy teenager's sudden decline on anything but the obvious culprit. And when the de Garays consulted area doctors for help, Stephanie says they too seemed oddly dismissive of Maddie's role as guinea pig for the Covid vaccine. *What's the purpose of the study if they're going to ignore the patients who get sick?* Stephanie wondered as she found herself reluctantly following advice to send Maddie to pain management specialists and a psychologist, who claimed the illnesses were all in Maddie's head.

"We did what they told us," Stephanie says. "I'm embarrassed how much I was brainwashed and believed them, but when they say, 'Hey, this is what's causing your daughter's symptoms, if you don't believe it, and you're not part of the solution, then she's not gonna get better.' So I did it."

Of course, the pain therapy and psychiatric counseling didn't make Maddie better. She got worse. By the time she was hospitalized for a third time, she was so sick she couldn't feel her legs, couldn't walk, had an abnormal heart rate, was suffering seizures, passing out, vomiting, unable to swallow or eat, and had to be fed through a tube. She was in the hospital that time for a month and a half.

Left to their own devices, Maddie's parents began doing research to try to figure out what was really wrong and how to help their daughter get better. In July 2022, after a year and a half of suffering, Maddie finally saw a neurologist specializing in vaccine reactions that cause neurological problems.

"She was diagnosed with chronic inflammatory demyelinating polyneuropathy," says Stephanie. "So basically, that's where your body attacks your nerves. . . . It's very similar to Guillain-Barré. So the treatment for that is IVIG, intravenous immunoglobulin, and she gets that every month." Guillain-Barré, a paralyzing nerve disorder, is a known reaction to flu shots and other vaccines.

After speaking with her mother, I briefly visit Maddie in the makeshift bedroom her parents have fashioned for her on the ground floor of their Cincinnati home because she can't climb the stairs to her own bedroom. The house was recently outfitted with an add-on wheelchair ramp. I'm preparing a report about Maddie's experience for *Full Measure*. I see a small figure lying in bed on her side, entirely covered in a sheet except for her eyes and nose, which are facing a sunny window. She doesn't feel like talking. She isn't feeling well after her latest intravenous immunoglobulin treatment. I tell Maddie I'm sorry for what's happened to her. I tell her that maybe my reporting about it can help find ways to help her get better faster, and it might help other people too.

Perhaps the biggest insult added to Maddie's injury is what the de Garay family sees as a lack of concern by the FDA. The study's leader did report Maddie's illness to Pfizer, which reported it to the FDA. But Stephanie says, "They left out a lot of information. . . . They reported it as 'functional abdominal pain.' So that's basically a stomachache that you can't explain." In reality, Maddie's condition was far more severe. She couldn't hold up her neck. She had no feeling from the waist down. She couldn't swallow or eat.

Family and advocates repeatedly alerted the FDA about Maddie's vaccine reaction, thinking that surely, if only they reached the right people, there would be help. But they continued to feel the cold shoulder from those they'd entrusted with their children's precious health.

Incredibly, neither Pfizer, Cincinnati Children's Hospital, nor the government will pay for Maddie's expensive treatments, so the family's savings are being depleted. As of this writing, Maddie is still using a wheelchair and a feeding tube. She can swallow small amounts of water and food, though it wreaks havoc with her stomach, so she's seeing a new specialist for that. She's getting some feeling back in her legs, can hold up her neck and head on her own, and can sit up. In these small advances, she and her mom find hope that she'll eventually recover. "I do think she will," says Stephanie. "I don't even know how to explain it—and she feels the same way—we've always felt like she will get better."

"What is important coming out of this that you think you would like people to know?" I ask Stephanie during our visit. This is a woman who had fiercely believed in and defended vaccines and the process of getting Covid vaccines into the arms of America's children.

"Well, first of all, you can't trust the government or hospitals or doctors," she says. "Doctors only know what they're told. They don't try to figure things out like you think. It's not like on TV shows. I've learned that this is much bigger than just the Covid vaccine. There's a lot more going on—it's scary. It's corrupt and all lies."

The story of Maddie de Garay's tragedy demonstrates how little doctors learn about these sorts of vaccine reactions in medical school or in Continuing Medical Education (CME) courses. When doctors are faced with the illnesses in real life, they typically don't recognize or report them, let alone know how to treat them effectively. As I look through the May 2023 CME course on Medscape, I see that it makes the claim that Covid vaccines are up to "100% effective" in kids. The fine print reveals the questionable way they arrived at that unbelievable metric. They followed the kids for just seven or fourteen days after vaccination. If the child didn't get confirmed Covid during those one to two weeks, the vaccine was deemed "effective." I kid you not. *In what world is a vaccine's performance for only one or two weeks indicative of anything about true effectiveness?* Furthermore, on Slide #20 claiming "100% effectiveness," the fine print at the bottom indicates that stat is from a study led by none other than Dr. Robert Frenck—from Maddie's study. In the paper he wrote up about his study, Dr. Frenck reported, "there were no vaccine-related serious adverse events and few overall severe adverse events."

For the purposes of educating other doctors, Maddie simply doesn't exist.

Another slide in the Continuing Medical Education slide show is titled: "Arguments for and Against Vaccinating Children for Covid-19." None of the "against" points reflects the most relevant facts: that vaccines don't prevent infection or transmission, that almost no children get seriously ill from Covid anyway, that natural immunity is probably superior and longer-lasting than vaccines, and that some children have died after vaccination. Instead, the presentation makes a passing mention of "myocarditis," or heart inflammation, then downplays it by calling it "rare and self-limiting."

It is a strange time, indeed, when doctors are trained to be incurious and incautious, and when they are trained in ideology over evidence.

Conflicted Teachers

Who is behind this biased presentation to doctors, I wonder, about the Continuing Medical Education (CME) class on Covid and kids, while intuitively knowing the answer. I take a close look at Slide #2. It shows names and photographs of the CME course teachers, followed by their long, distinguished titles. There's Saul Faust, Professor of Paediatric Immunology and Infectious Diseases, and Director of a Clinical Research Facility at the University Hospital Southampton in the UK. There's Dr. George Kassianos, National Immunization Lead at Royal College of General Practitioners, and President of the British Global and Travel Health Association. And there's Dr. Barbara Rath, head of the Vaccine Safety Initiative in Berlin, Germany, and Research Director at a French university. What's left off Slide #2, though, is the esteemed faculty's financial ties to drug companies.

"All relevant financial relationships for anyone with the ability to control the content of this educational activity . . . have been mitigated," claims Medscape on its web page about the course, though it's unclear what "mitigated" means. I click on the disclosures under the names of the instructors and finally get the answer to the question of who's paying them. Dr. Kassianos advises or consults for Merck, Novavax, Pfizer, Sanofi, Takeda, and Valneva. Dr. Rath of the "Vaccine Safety Initiative"? She's a consultant or advisor for AstraZeneca, GlaxoSmith-Kline, and Roche. For Faust, he's got relationships with an A-to-V list of vaccine manufacturers: AstraZeneca, GlaxoSmithKline, Johnson & Johnson, Moderna, Novavax, Pfizer, Sanofi, and Valneva. One might think it would be crucial for doctors who are taking the course to know about the business relationships of their instructors. But that's not presented on the slides.

I note another issue with the financial disclosure requirements for instructors: they only cover the previous two years. In other words, if a CME instructor got $50,000 from a vaccine maker two years *and one day* earlier, it's deemed irrelevant. Two years is an unreasonably short time period for disclosure requirements. It exempts instructors from having to fess up to a lot of potential conflicts of interest.

Of course, here I am dissecting the fine points of what and how disclosures should be made, when the practical position is that paid experts shouldn't be designing or teaching medical courses about their companies' products. It's difficult to find justification for why the medical profession permits this. If you were trying to decide whether eating pesticide-laden food is harmful, would you want to get all your facts from people working in the pesticide industry? If you wondered whether water was safe to drink in East Palestine, Ohio, after the toxic Norfolk Southern train derailment, would you gather all of your information from people paid by Norfolk Southern? If you want to know if a medicine is safe, should all of your data come from the company making the drug?

Yet, this is the status quo in medicine and science. There are so many experts taking money from the pharmaceutical industry that those in charge say it's unrealistic to ban all of them from advising on medical issues. So, instead of admitting it's a problem that's grown out of control, they simply adjust their policies and moral compasses accordingly. These same people argue with a straight face that it makes perfect sense to rely on industry-paid experts because *they're* the ones who know the most. *See? It's a perfectly logical approach. In fact, it's the best!*

From designing and teaching medical courses, to advising physicians, to sitting on government and professional expert boards, relying on corporate-paid medical experts is a ubiquitous but ethically questionable practice. It makes it difficult for safety concerns to be addressed in a fair and thorough manner that benefits patients.

Turning a Phrase

There are countless additional ways in which the scientific profession seamlessly advances the pharmaceutical agenda. With experience and hindsight, I've come to recognize propaganda strategies behind the use of medical phrases that I once accepted without question. I'll identify several examples to help guide you on your own quest to divine the truth amid a sea of medical misinformation.

The first example is advancement of the notion that an illness caused by a medicine isn't real—unless it's been identified through a "**peer-reviewed, published study**." In fact, powerful indicators of side effects can be found among anecdotes and clinical observations much sooner than we have those peer-reviewed studies. And, as we have seen with manipulation of studies and study data, even those peer-reviewed studies can be dubious.

I initially recall hearing "Where's the peer-reviewed, published study?" spat at me by a *60 Minutes II* producer at CBS News who wanted to kill a news story. He'd been assigned to work with me on a topic that he clearly didn't want to report. The story was about an illegal immigrant who had enlisted in the US military, but upon receiving a battery of vaccinations had become seriously ill and died. The Army and the country he'd served abdicated all responsibility. His family was left with lingering questions.

The story I intended to report would not debate the side effects of vaccines so much as tell a larger, nuanced story about the death of a deeply patriotic man who was forsaken by the country he'd served, due to his illegal status. This was at a time when pharmaceutical industry influence was taking hold in America's newsrooms and I later came to realize that most any story that mentioned or exposed vaccine safety issues was considered a nonstarter.

It was also becoming clear that some insiders at CBS were fully on board with the propagandists. To this day, I don't know if it's because they were somehow compensated—which is what their unjournalistic behavior suggested—or were simply susceptible to the clever disinformation dispatched by misinformationists. Whatever the case, I noticed they responded to stories reporting on drug safety, particularly vaccine injuries, with a visceral, disgusted repulsion that defies logic and normal editorial judgment.

As I'm on the telephone outlining elements of the planned story on the illegal immigrant, the *60 Minutes II* producer interrupts. "Where's your peer-reviewed, published study?" he growls in an argumentative tone. At first I'm not sure what he means. Obviously, there would be no "peer-reviewed, published study" on this singular man's recent death.

"Study on what?" I ask.

"Proving that vaccines caused the guy's death," he replies. "Where's your peer-reviewed, published study?"

Indeed, in a more general sense, there are plenty of scientific articles and court cases documenting the types of injuries and deaths vaccines can cause. As I begin to explain, the producer angrily ends the call. At the time, CBS is what we call a "strong producer shop." This means there is almost no way for a young reporter to get a story on the air unless a producer is on board with it. So that was the end of that.

I more frequently began to run into the same syndrome: a medical side effect disregarded as a myth unless it was documented in a "peer-reviewed, published study." And it made less and less sense.

I reflected on the fact that when my own reporting broke news on medical safety issues over the years, it was never due to a peer-reviewed, published study. Instead, evidence had percolated from the ground up, almost in spite of mainstream medicine's seeming desire to bury it. Some safety issues get exposed by whistleblowers and other insiders at corporations or government agencies. Some are revealed by independent doctors who notice trends and report case studies. Some are unearthed by patients doing their own research and connecting with other injured patients online. Issues are uncovered through reports to federal adverse event databases that can be searched for clues and trends, by attorneys representing the injured who obtain revealing documents and data in the legal discovery process, by watchdog groups conducting deep-dive research, and so on. All of those can provide compelling evidence. In fact, in my experience, one of the slowest and least reliable methods to discover safety concerns is to wait for them to appear in "peer-reviewed, published studies."

I came to believe the "peer-reviewed, published" argument was invented by industry propagandists; adopted by federal agencies; pushed out to doctors, scientists, and public health officials; and parroted by medical reporters, TV doctors, and others in media. Convincing the public that a peer-reviewed, published study is required before an adverse event can be considered "real" guarantees there will be serious lag time before action is taken that stands to hurt a pharmaceutical company's bottom line. Convincing the public that a peer-reviewed, published study is required to validate an alternative medical treatment, one that doesn't involve a big-selling pharmaceutical product, guarantees that the big-selling pharmaceutical product will remain the go-to choice for most doctors and insurers. Peer-reviewed, published studies take time—when they're launched at all. The biggest funders of peer-reviewed, published studies are government and the pharmaceutical industry, and the likelihood that they'll publish studies casting doubt on their star moneymaking commodities is low. Convincing us to wait for the study that may never come allows a company to successfully forestall warnings or forced withdrawal of their product for months or years.

Delving deeper into the rabbit hole, the entire idea that there's something magically validating about the much-heralded peer-reviewed, published study falls apart. The credibility of the study depends, in large part, on the peer reviewers.

A contemporary example of that lies in an email exchange from the early days of Covid on February 16, 2020. The documents were later obtained by the House Select Subcommittee on the Coronavirus Pandemic. The emails involve key researchers discussing an article they intend to publish to convince the public that Covid "came from Nature, not a lab." One of the researchers is a Tulane professor named Robert Garry. He refers to the questionable practice of researchers being permitted to cherry-pick peer reviewers who will approve the scientific article for publication. Garry tells a coauthor, Professor Edward Holmes of the University of Sydney, "So, as you know when you submit [our article for publication] you'll need to suggest [peer] reviewers to include and exclude. Seems easy—there are some natural choices for both lists." To which Professor Holmes replies: "Oh yes, the reviewers are easy . . . I think this is a slam dunk." Their article is published on March 17, 2020, in *Nature* on an impressively quick timeline, just a month after the email discussion. It's entitled, "The proximal origin of SARS-CoV-2." It's circulated by propagandists in government and media to "debunk" the "lab origin theory."

Investigative medical reporter Paul Thacker later sarcastically summarizes what the email exchange implies. On Twitter, he writes, "2 veteran researchers discuss how scientists publish in 'peer reviewed' journals: YOU pick your peer reviewers. The much lauded peer review process uncovered. This is how the magic happens, baby."

Without getting too far afield, I'll mention a familiar figure is copied on the referenced email exchange about choosing which peers would review an article attempting to debunk Covid's lab origins: Professor Ian Lipkin of Columbia University. It's the same Ian Lipkin who had, years earlier, assured Hollywood star Robert De Niro that vaccines can't possibly cause autism. This time, he's coauthor of the study attempting to undermine any notion that Covid came from a Chinese lab. The lesson here is that when propagandists hold positions of influence and are not held accountable for their mistakes or dishonesty, they continue to misshape public opinion and science in one public health scandal after another on behalf of the pharmaceutical industry.

So today, when I hear an authority advance the idea that a medical risk is supposedly unproven because there's no "peer-reviewed, published study," you can understand why I tend to be skeptical. Yet the strategy seems to work well among an unthinking media. *There's no peer-reviewed, published study*, a medical authority figure tells a journalist, trying to throw him off the scent of a given medical scandal. The authority raises an eyebrow for emphasis, tilts his head back a little, and looks down his nose. *Of course,* nods the journalist knowingly, without knowing a thing at all. *There's no peer-reviewed, published study.*

There are several other phrases in the same general category as "peer-reviewed, published study." One of them is "**Randomized Controlled Trials**," as in *Where's your Randomized Controlled Trial (RCT)?* One observer I agree with refers to the concept of RCTs and their supporters as, "The temple of . . . Randomised Controlled Trials (RCTs) and their worshippers." He states that it's improper when doctors disregard treatments because they haven't been through Randomized Controlled Trials or meta-analysis. He points out that some questions don't need Randomized Controlled Trials for verification, or that some such studies can't be done due to the "rarity of the conditions or ethical issues." In addition, "Well-conducted RCTs are often expensive, labour-intensive and take time to perform and reach their conclusion." Another possible stall tactic.

A close cousin to those phrases is "**No proven causal link.**" *Coincidence isn't causation,* medical authorities are fond of saying about possible drug side effects. *There's no proven causal link.* They may toss in an example, such as, *I drink coffee every morning, and the sun rises. But does drinking coffee make the sun rise?*

While that's obviously true, it's also true that if every time you drink coffee you get a stomachache that you never get any other time, it's reasonable to ask if the coffee might be the cause.

In a medical context, drugmakers may spin the fact that a lot of people seem to get liver damage after taking a certain drug. *There's no proof it's "causal." After all, there's no peer-reviewed, published study showing that the medicine could actually "cause" liver damage. And until then, it's just a conspiracy theory.*

Yes, nods the journalist knowingly, without knowing a thing. *There's no proven causal link. It's a conspiracy theory. Debunked. If you keep asking questions, you must be a conspiracy nut!*

About fifteen years ago, the government secretly paid a multimillion-dollar settlement for autism and other vaccine damage to young Hannah Poling, then sealed the case to keep it secret. When the settlement leaked out to the press, the government spun the story like a top. *Okay, yes, maybe we paid for Hannah's vaccine injuries. But vaccines* triggered *her autism. They didn't* cause *her autism. See? The link isn't causal. So vaccines don't "cause" autism.*

It takes no more than a modicum of critical thinking to understand that the difference between vaccines "triggering" or "causing" an illness may be meaningless—at least to the injured. For example, scientists say children who have predispositions for vaccine injury can avoid triggering the injury and live perfectly healthy lives. It would simply require what scientists like Dr. Bernadine Healy once suggested is entirely possible: identify the predispositions, then vaccinate those kids differently, develop different formulations for them, or avoid some of the most problematic vaccines entirely.

Another term to watch out for is something doctors refer to as "**Evidence-Based Medicine**," or EBM. In the 1990s, EBM was widely established and adopted. It means that patients should be treated based on the "current best evidence." Sounds reasonable. The problem is: *Whose* evidence? As we've seen, what does and doesn't get published in scientific journals is largely controlled by special interests. Furthermore, evidence typically builds and changes over time.

In a letter published in the *British Medical Journal*, an orthopedic surgeon rightly points to many reasons why EBM is flawed. He says the "blind pursuit of the holy grail of statistical significance" in studies can result in doctors overlooking harm to patients as well as ways to help them. Evidence-Based Medicine guidelines can be biased due to institutional support and conflicts of interest among those determining what constitutes EBM. The guidelines can be "slow and out of date."

"It is not unusual for various national clinical networks to take 5+ years to form a consensus which itself is soon overtaken by new revelations and technology," writes the surgeon. He adds that results of EBM studies may be consciously or unconsciously biased by the researchers. And finally, another flaw with Evidence-Based Medicine is that "no evidence of an effect" isn't the same as "having no effect." Put another way, just because evidence of a benefit doesn't show up in a particular study, it doesn't prove there's no benefit. Many doctors confuse the two. A particular therapy or treatment may be the best current option for an individual patient even if there aren't high-quality studies showing that it works. It may be a starting point to gain further evidence one way or another. Yet doctors seem to be taught to repeat, as if a memorized chant, that they subscribe to the practice of "Evidence-Based Medicine" and so cannot try something new that might help a patient, or won't halt a treatment that seems, anecdotally, to be causing harm.

You may also hear propagandists dismiss studies they don't like by claiming that "**the bulk of the science**" or "**the bulk of the evidence**" indicates something different. Or they may refer to "**the scientific consensus**." These phrases are trotted out when a journalist or injured patient manages to satisfy the established burden of proof and show there are, indeed, peer-reviewed, published studies—Evidence-Based Medicine—that support a claim about a medical side effect. A shift in argument is called for. *Okay,* say the propagandists, *maybe there are peer-reviewed, published studies on that. But* those *studies aren't true. The "bulk of the science" comes down on the other side.* Or *"the scientific consensus" says otherwise.*

"The scientific consensus" argument should raise suspicion each time you hear it. It's presented as if it's irrefutable. Yet, the scientific consensus is wrong *All. The. Time.* Until the 1970s, doctors didn't think cholera could stay alive in water without human hosts, until one scientist shocked them by showing it did. Shoe-fitting x-ray devices were found in shoe stores through the 1950s, with the consensus being it was a safe, fun way to measure foot size—until someone finally discovered the radiation was harmful. Until 1961, doctors in Europe were happily dishing out thalidomide to pregnant women for morning sickness, until it was discovered to be responsible for horrifying birth defects. Think of the many drugs and vaccines, some of them blockbusters in terms of sales, that the scientific consensus deemed to be safe and effective, but which were later pulled from the market for being unsafe or ineffective. Medical history is littered with examples of the scientific consensus being mistaken on important facts, and ganging up against a lone scientist or two who turned out to be correct.

The first time I heard the phrase "bulk of the science" was when I was assigned to cover vaccine safety issues at CBS News. I routinely sent my stories to the CBS legal department for review prior to air. The stories were not legally precarious, and the legal reviews were never difficult. I considered them to be an additional layer of protection for me and my reporting for when the pharmaceutical industry's global law firms would go on the attack. This line of stories always drew backlash from vaccine industry interests looking to undermine them and smear me. Over time, I learned what the CBS lawyers looked for in my news stories in terms of wording and supporting evidence and it helped me frame my reporting in a legally defensible, fair, and accurate way.

Around this time, as I've mentioned, the pharmaceutical industry was growing heavy-handed and more influential with news organizations. One day, a CBS attorney reviewing my story about vaccines and autism suggested we add a phrase: "The bulk of science finds no link between vaccines and autism." At the time, I figured it was a minor addition and no harm done, so I agreed without giving it much thought. *Yes,* I thought, knowingly, without knowing a thing. *Certainly the bulk of the science finds no link.*

Later, when I reported my next story on the subject, the CBS attorney again suggested I add the same line. Clearly, I came to believe, there had been a discussion about these stories at some level within the CBS organization. I knew that pharmaceutical advertisers were lobbying CBS corporate folks against me and my stories, and I suspect they had suggested inserting the invented phrase if the stories could not be stopped entirely. But this time, I'd given it some critical thought. *Was it really the case that "the bulk of the science" found no link between vaccines and autism?* I hadn't personally conducted a review of "the bulk of the science," which likely amounted to thousands of studies. In fact, I didn't know of anyone who had. Conversely, the bulk of the science that I was personally familiar with suggested vaccines are, indeed, linked to autism. So I told the CBS lawyer, "I just don't know for sure, and I haven't seen anybody independent do an analysis, but from what I know, the bulk of the science does suggest a link between vaccines and autism. So if we want to say in our story that 'the government and vaccine industry claim the bulk of the science rejects a vaccine-autism link,' that's fine, but I can't personally state it as if I know it to be true."

Soon, "the bulk of the science" phrase needed to evolve into a slightly amended form. It turns out a few journalists actually did their homework and learned that "the bulk of the science" *does* support some inconvenient fact, such as the link between a medicine and a side effect. So now, the propagandists say, "the bulk of the *credible* science finds no link . . ." *Ah*, say the ignorant reporters without knowing a thing, *credible* science *finds no link*. Of course, the supposed "credible" studies are the ones funded by the company that makes the medicine.

Today, I hear powerful interests routinely spout these phrases when they need to discredit a scientific idea. Propagandists deploy the terms. Government agents, public health officials, journalists, doctors, and analysts in the media, who don't care to scratch beyond the surface, amplify them. The words are weaponized to distract and deflect.

Medical School Secrets

Another way the pharmaceutical industry plays a pivotal role in influencing what America's doctors believe and how they treat us is through medical school.

In the mid-2000s, an acquaintance who's a doctor and university professor was excited to tell me he'd accepted a position as an administrator at a medical school. He said his main goal was to try to banish drug industry influence. He was anxious to accomplish something necessary and meaningful. But after a year on the job, he told me he'd given up.

"Nobody in the [university] administration supports reducing pharmaceutical influence," he told me with frustration bordering on despair. "I couldn't even get them to agree on the smallest steps, like stopping companies from bringing in lunch for students. They fought that tooth and nail. I'd never be able to tackle the big conflicts."

Beyond the free lunches, what are some of the "big conflicts"? For one, how about the incredible fact that some of the biggest reference books med students and professionals are taught from are actually published by a pharmaceutical company! *The Merck Manual of Diagnosis and Therapy* is referred to as "the world's bestselling medical textbook." One reviewer calls it "[t]he most basic book that I used to survive 1st year to even 3rd year med school." There's also a handy, dandy, free consumer version. Merck advertises its Manuals (there are several editions) as "the best first place to go for medical information" and writes they are "one of the world's most widely used medical information resources . . . committed to making the best current medical information accessible to healthcare professionals and patients on every continent." And the Manuals are provided for free. *Imagine that! How selfless of Merck!*

It amounts to a massive conflict of interest. But Merck would say otherwise. In its Manuals, Merck writes, "The US Food and Drug Administration requires Merck Manuals to maintain a strict separation between the publications and Merck & Co. to avoid any bias toward drugs produced by Merck." Indeed, Merck points to various steps it takes to remove any pesky concerns about conflict of interest. An "independent" medical board reviews the articles, authors cannot be employed by Merck, and medical articles are set apart from commentary and news. Merck also promises, "[a]lthough the editorial staff is employed by Merck & Co. . . . there is no control, review, or even input into the content of the Manuals allowed from any other part of our company." So the Merck-employed publishers are pinky-promising that they can be perfectly trusted to police themselves, and nobody could possibly be sneaking in any bias on behalf of the company or pharmaceutical industry.

While it would be nice to take Merck at its word, it's not difficult to find reason to be skeptical. Merck has not always proven trustworthy. For example, in 2011, Merck agreed to shell out $950 million for dishonest and unethical behavior involving Vioxx. Vioxx was a painkiller pulled from the market in 2004 after a long-running controversy over its dangers. Merck was said to have made false, inaccurate, unsupported, or misleading statements about Vioxx's heart safety in order to boost sales. Penalties included a $321.6 million criminal fine for illegal promotion and marketing of Vioxx. Do Merck's Manuals tell doctors and med students about its own sordid history of deception and fraud? Apparently not. An online search of "Vioxx" in both the professional and consumer version of the Merck Manuals turned up no results on this score.

Do Merck Manuals mention that the company's own HPV cervical cancer vaccine, Gardasil, has been the center of major controversies about its safety and effectiveness? Does it disclose that injured patients have filed many lawsuits claiming the vaccines caused illnesses from ovarian failure to cancer? That the scientist who codeveloped Gardasil later spoke out in an unprecedented way, saying that the Merck Gardasil vaccine may have more risks than benefits? No. Instead, Merck's Manual makes an audacious claim under "Side Effects of HPV Vaccine." It states flatly and falsely, "No serious side effects have been reported." It's unknown how that claim could possibly square with Gardasil's FDA-approved label, also written by Merck, which states: "the following postmarketing adverse experiences have been spontaneously reported for GARDASIL: Blood and lymphatic system disorders: Autoimmune hemolytic anemia, idiopathic thrombocytopenic purpura, lymphadenopathy. Respiratory, thoracic, and mediastinal disorders: Pulmonary embolus. Gastrointestinal disorders: Pancreatitis. General disorders and administration site conditions: Asthenia, chills, death, malaise.

Immune system disorders: Autoimmune diseases, hypersensitivity reactions including anaphylactic/anaphylactoid reactions, bronchospasm. Musculoskeletal and connective tissue disorders: Arthralgia, myalgia. Nervous system disorders: Acute disseminated encephalomyelitis, Guillain-Barré syndrome, motor neuron disease, paralysis, seizures, transverse myelitis. Infections and infestations: Cellulitis. Vascular disorders: Deep venous thrombosis."

Blood clots, paralysis, seizures, brain damage, and death—yet the Merck Manual online tells med students, doctors, and consumers, "No serious side effects have been reported"? And how about this little beauty included on the information label for Gardasil 9: Gardasil isn't recommended for pregnant women but 62 test subjects got pregnant 30 days before or 30 days after vaccination, and 18 of the pregnancies did not end with a live birth. There was an astounding 27.4 percent miscarriage rate, which is more than double that of women given a different version of the shot.

An unbiased textbook would include a fair recitation of these cons as well as the pros, and urge doctors to monitor their patients for possible side effects. Instead, Merck irresponsibly teaches doctors that serious side effects simply do not exist.

The problem extends far beyond the Merck Manuals. A 2022 study examined nine textbooks commonly used to teach which medicine to prescribe for various psychiatric disorders. Two-thirds of the textbook authors and editors had been personally paid by companies that make the drugs.

Naturally, I was curious about how Merck Manuals treat the autism and ADD epidemics that—as of this writing—afflict more than 1 in 36 eight-year-olds in the US. Is there a proportional sense of alarm sounded over the skyrocketing numbers and the medical community's inability to slow them? Does Merck, the maker of a measles, mumps, rubella (MMR) vaccine, tell the story of the CDC senior scientist, Dr. William Thompson, who became a whistleblower and told Congress that he and his colleagues altered a study to minimize links between MMR vaccine and autism in black boys? Do the Merck Manuals include the opinion of the government's own pro-vaccine expert, renowned pediatric neurologist Dr. Andrew Zimmerman, who came to conclude that vaccines can cause autism, after all? And who went on to swear under oath that Department of Justice attorneys fired him as their expert witness when he told them about the connection, and that they then covered up his opinion in court? Do Merck Manuals discuss the numerous federal vaccine court payments for children injured by vaccines, who ended up with autism? Are vaccine-related brain injuries examined? Is there mention of the thousands of studies that implicate vaccines in a host of disorders, from autism to chronic immune disorders that are so disturbingly common today? Does Merck disclose to fledgling doctors that "autism" is listed on the label of the since-discontinued Tripedia (DTaP) vaccine under "adverse events reported" (along with "idiopathic thrombocytopenic purpura, SIDS, anaphylactic reaction, cellulitis, autism, convulsion/grand mal convulsion, encephalopathy, hypotonia, neuropathy, somnolence and apnea")?

My online search of the Merck Manuals for relevant keywords related to these topics didn't return any of those disclosures. When I checked under the heading of "autism," the Merck Manual simply states, "The cause in most children is unknown, although evidence supports a genetic component; in some patients, the disorders may be associated with a medical condition. Environmental causes have been suspected but are unproved. There is strong evidence that vaccinations do not cause autism, and the primary study that suggested this association was withdrawn because its author falsified data." The misinformation contained in this paragraph addressing the biggest and most long-standing epidemic among America's children is breathtaking. *Nope. No bias there at all.*

Yet this medical publication is relied upon by medical professionals who hold our lives in their hands. A pediatrician once told me that among colleagues, the Merck Manual is commonly referred to as "the pediatric Bible." Not "the pediatric encyclopedia," but the "Bible"—as if to be followed religiously, without question, without thinking.

CHAPTER 15

New Warning Signs

The worst, newest health crisis exacerbated by manipulation of our information in the name of science is already impacting millions. It has to do with illnesses loosely grouped under "Long Covid" and "Long Vax." Based on what researchers are learning, they might more accurately be characterized as "spike protein" and "microclot" disorders.

By February 2024, even the medical establishment admitted that an alarming number of adults, up to 25 percent, shows signs of "Long Covid" illnesses. So far. Not surprisingly, nearly every public account omits an obvious and critical fact: most people who have "Long Covid" were vaccinated, so their symptoms may, in fact, be due to—or worsened by—the vaccines.

Whatever the syndrome is named, it amounts to yet another national health emergency that's not being treated like one. There is no cohesive action plan. No well-publicized effort to develop treatments or protocols to identify and help afflicted patients. And with all the billions of taxpayer dollars committed to various Covid-related efforts, there's no major initiative to educate doctors in real time so they can help their patients. In fact, if you hear anything at all in the media and medical journals, it tends to amount to spin. They pump up the dangers of "Long Covid" to convince more people to get vaccinated over and over, leaving out the warning that each dose of vaccine may worsen the risk.

Independent researchers are learning that these Long Covid and Long Vax illnesses can surface long after the virus clears the body. The strange spike protein in Covid, and made by the vaccines, sticks around and apparently has a plan. It takes hold of people even when they may not have felt sick during Covid or right after vaccination. It can suddenly manifest months or years later and invite a cascade of inflammation, immunity, and blood problems. The result is a diverse range of symptoms and illnesses that exploit a person's innate weaknesses, ranging from retina detachment, rigid muscles, headaches, gastric issues, hearing problems, blurred vision, rashes, sore muscles, fatigue, weakness, shortness of breath, arthritis, bone weakness, and brain fog; to fainting, autoimmune disorders, heart attack, collapsed iliac vein, paralysis, atrial fibrillation, nerve damage, and stroke. According to scientists, the spike protein made in your body by the vaccines can prove even more debilitating than the one in Covid. And two doctors who are treating thousands of such patients tell me the sickest ones they see tend to be those who got the most vaccines and the most cases of Covid—which are usually one and the same patient. If leading-edge researchers are correct, Long Covid and Long Vax will strike nearly all of us in some form—if not now, then when our immune systems weaken with age or illness.

So your average physicians get left in the dark, without test and treatment guidance from the government or professional medical associations. They're telling sick patients that there's nothing wrong with them, or that their routine tests are normal, or that their problems are due to anxiety or all in their head, or maybe that they have multiple sclerosis. As a rule, these doctors lack the time or curiosity to do the deep dive necessary to understand the worsening crisis and how to address it.

A brave, independent few are stepping up as medical detectives in a scientific environment that's arguably more uncharted than the AIDS crisis in the early 1980s, and impacts far more people. These sleuths are working to figure things out on their own, without the benefit of federal funding and despite establishment forces trying to smear them or threatening to pull their licenses.

Early one fall morning in 2023, I'm boarding a connecting flight to Birmingham, Alabama, to meet one of them.

The Clot Factor

Dr. Jordan Vaughn is an internal medicine specialist and CEO of MedHelp Clinics, an independent group of treatment centers. When I arrive with my television crew for interviews in fall of 2023, the main clinic is doing a bustling business. Some staffers show us around and offer southern hospitality in the form of Alabama peanuts, cheese and sausage balls, and other snacks.

Dr. Vaughn arrives and greets me with an enthusiastic handshake. He's wearing a checked button-down shirt and gray pants. A stethoscope sticks out of one oversize pocket on his white lab coat. He has brown hair, a receding hairline, and a friendly smile. He speaks quickly and has a lot to say, enjoying the act of sharing the knowledge he's gained. While we wait for patients to arrive to tell me about their experiences, he continues conversations we'd begun on the phone and via text. To me, it seems like Dr. Vaughn never rests. He's always thinking about ways to tackle the vexing new challenges: Long Covid, Long Vax, microclots, and the spike protein making so many people sick.

Among the first patients to arrive at the clinic to talk with me is Vandiver Chaplin. If there's a singular person who originally set the gears turning for Dr. Vaughn, it's Chaplin. It was in December 2020, shortly after Chaplin got his Covid vaccines.

"I just felt terrible," Chaplin tells me. "You know, dizzy, lethargic, all those kinds of things. I was having some optical issues too. My vision would just go blurry suddenly, and then maybe a minute or two later, it would clear up."

Chaplin goes to say that Dr. Vaughn, his longtime physician, started the diagnostic process by running some scans and other tests. But nothing abnormal showed up. Then, considering that Chaplin had significant shortness of breath, which can be a symptom of blood clots, Dr. Vaughn decided to do some special blood work. He hit upon something crucial.

"I found that he had abnormal clotting issues," Dr. Vaughn says. The abnormal blood-clotting issues were there all along, just hidden from view—so tiny, they couldn't be detected on regular scans. "So I treated him as if he had something that I wasn't able to totally see, which would be smaller vascular issues, and his symptoms significantly improved. So that really pushed me off on a, really a kind of a journey to say, 'What is going on here?'"

As you may be able to tell, Dr. Vaughn is that rare breed of physician who asks questions and tries to figure things out rather than dismiss people's complaints. Obviously, that makes him very popular among patients. By the time of my visit, he and his team had treated more than 1,100 patients for Long Covid or Long Vax, from athletes in their teens to people pushing age ninety. Their symptoms were as diverse as their demographics. Some became sick right after Covid, after Covid vaccines, or both. Others were hit hard a year or two later.

Thirty-nine-year-old Hannah Bourgeois and her husband, Dr. Greg Bourgeois, were both vaccinated and got Covid. Hannah became short of breath and eventually got so sick, she was nearly bedridden for two years.

"I felt like my body was just shutting down on me, and there wasn't anything I could do about it," Hannah tells me. After a consult with the famed Mayo Clinic brought her no improvement, Greg, a dermatologist who attended medical school with Dr. Vaughn, heard about what Dr. Vaughn was doing and sought him out to treat his wife.

"So I learned that there were a lot of microclots kind of throughout my body that was just causing oxygen not to be able to get around very well," Hannah says, still looking thin and frail, though on the path to recovery. "He was the first doctor that when I went to see him, he would finish my sentences for how I was feeling. . . . I think I started to cry the first time because that was so new, and he understood, and he said, 'You know, it all makes sense.'"

"What is the treatment he gave you, and how do you feel today?" I ask.

"So he put me on the triple anticoagulant therapy, and within a couple days I started to notice some difference. But within two weeks,

I felt like I had risen from the dead. I mean, I got my voice back. I could walk. I could do things."

Another physician who sought Dr. Vaughn's help is Dr. Donald Carmichael, age eighty-eight. He's a retired vascular surgeon and former professor of surgery at the University of Alabama at Birmingham. He and his wife, Mary Alice, both vaccinated and boosted, say they got Covid more than once. The last time was a near-killer for him.

"We thought he was not going to live through the night," says Mary Alice. "Our son, who is a friend of Dr. Jordan Vaughn, said, 'We are not taking you anywhere else except to him in the morning.' His treatment put him back in, basically in full health. And he was so giddy, I thought he had lost his mind!"

Dr. Vaughn is also treating young athletes. The case of fifteen-year-old Braden Little baffled doctors for two years. More than once, Braden suddenly collapsed on the basketball court after he had Covid. After starting Dr. Vaughn's treatment, Braden became so vastly improved that he was back on the road shooting hoops again when I visited Dr. Vaughn's practice in Birmingham.

Nineteen-year-old runner Ellen Redinger is just beginning her journey with Dr. Vaughn when we meet and tells me she hopes for a similar recovery. She was vaccinated, got Covid, and became very sick.

"I mean, I can pinpoint the day, the time, where I was, when I was running. I was doing a workout. And all of a sudden, I cannot feel my legs . . . I mean, my heart rate is going 200. I can't do it. I call my dad. I'm like, 'I'm done working out. I can't, I can't do it.' And I went for like three or four months of just feeling awful," Redinger tells me.

Together, Dr. Vaughn and his small team had managed to begin unraveling some of the emerging mysteries that have become taboo to discuss in establishment medical practices, especially when it crosses into vaccine adverse events. Their work is centered on the spike protein in Covid and made by vaccines.

"I always say it's almost like there's two worlds. There's before Covid and after Covid," Dr. Vaughn says. "And a lot of doctors are still living in the 'before Covid' world, where everything's in the textbook. But when you have a syndrome that comes before you, and it happens to be associated with this new pathogen that everyone seems to have been in contact with, you've got to kind of open your eyes, open your ears, and also get into the literature and try to figure out what the heck's going on."

Numerous patients treated by Dr. Vaughn had been told by other doctors that their aches and pains were all in their head, or simply too mysterious to effectively treat. Some were told they might have MS. Some were sent for psychiatric help. But Dr. Vaughn listened and began to solve the seemingly insoluble puzzles. He thinks he's figured out why people who have had both Covid and Covid vaccines often seem to get the sickest. And it has to do with something called fibrin.

"So we are designed all to make fibrin. Fibrin is one of the first kind of response mechanisms," Dr. Vaughn explains.

"It forms a clot if you're injured or something?" I ask.

"Yeah . . . trauma or infection, all those kind of things. You're going to make fibrin as a response to that. The fibrin you usually make is like spaghetti that just came out of the colander. But the fibrin you make in response to a spike protein that's associated with Covid and the vaccine is kind of like burnt spaghetti with cheese in it that you have to get a Brillo pad and get it off the bottom of a casserole dish with. And in that sense, that's why it's so unique. It's resistant to being broken down. Literally everyone, when they have the spike protein exposure from either the vaccine or from the infection, you're going to make some of these amyloid fibrin. The question is, who can get rid of them? And if you can't get rid of them, they sludge up the small vessels and inhibit the delivery of substrates, and those are things like red blood cells, which carry oxygen. And so in that case, if you can't get oxygen to tissues, you're going to have significant dysfunction at every level."

While Dr. Vaughn is meeting with great success treating his patients, some published research is beginning to nibble around the edges of the problems. Yale researchers recently reported, "Persistent symptoms after vaccination 'long vax' are similar to those reported with long COVID." *Science* magazine writes, "Rare link between coronavirus vaccines and Long Covid–like illness starts to gain acceptance." Why hadn't the top public health experts and government scientists uncovered this on the front end so we could better understand the risks of vaccination and, if vaccination still seemed viable, be prepared to treat resulting complications?

Apparently, some government experts did learn about the illnesses being suffered and even treated a select few of them, but kept it a closely held secret. In the 2023 documentary *The Unseen Crisis*, Brianne Dressen speaks out. She took part in the AstraZeneca Covid vaccine trials. She got so sick after vaccination, she said she was sure she was dying. But nobody could diagnose her. Finally, she stumbled upon help. The National Institutes of Health flew her to the DC area to be part of a small study with twenty-two other vaccine-injured people.

"They flew us out. We were there in person," Dressen says. "We were in the state-of-the-art facilities. They know at a very intimate level what's going on with this. They know about microclotting. They know about the nervous system breakdown. They know about small-fiber neuropathy. They know about the cognitive issues. They know all of it. They haven't given that very essential and matter-of-fact treatment to all of the other Americans who stepped up and got their shot."

Cynthia Drukier, the producer of the film Dressen appeared in, tells me, "[The medical experts at NIH] studied it and they treated her. And they reversed the trajectory of her illness. And she got better, a lot better, almost completely better. And this was very early on." Yet to this day, the government isn't sharing what it knows. *Why?*

While researching a story on all of this for *Full Measure*, I'm looking over story tips sent to me by observers who send me emails or leave comments on my website SharylAttkisson.com. An interesting one comes from a friend, a nurse who followed the original FDA meetings that led to approval of the Covid vaccines. It turns out these very problems were both foreseeable and foreseen.

Premonition

Back in December 2020, a chilling foreshadowing had been given by a Harvard-affiliated pediatric specialist, Dr. Patrick Whelan. He'd already seen evidence in the literature that the spike protein made by Covid vaccines could cause great harm. In a letter to the FDA just before the first vaccines hit the market, Dr. Whelan wrote that if people catch Covid by accident, that's unavoidable, but affirmatively giving them the spike protein through vaccines is quite another matter that could prove devastating to the population.

"[I]t appears that the viral spike protein [created after Covid] vaccines is also one of the key agents causing the damage to distant organs," he wrote.

He acknowledged that if a vaccine were to work in preventing Covid, it would be a welcome development, but warned "it would be vastly worse if hundreds of millions of people were to suffer long-lasting or even permanent damage to their brain or heart microvasculature [small vessels] as an unintended effect of vaccines."

In summary, he said more study was needed before the FDA should consider allowing the vaccines to hit the market.

Dr. Whelan's pleas apparently fell on deaf ears, demonstrating, yet again, that checks and balances to ensure our medicine is safe are hopelessly compromised.

Thankfully, Dr. Vaughn forges on. He and his team continue to get swamped by requests for help from suffering patients as far away as Germany. He's also conducting original research and sharing what he's learning. He regularly speaks to groups of several hundred physicians, and holds roundtables at night on Zoom for doctors who are desperately looking to help patients but have reached dead ends looking to government advisors, public health officials, and their professional medical associations.

Changing Tide?

In February 2024, I'm in Phoenix, Arizona, reporting on a conference of doctors and other medical professionals. It's titled "Healthcare Revolution: Restoring the Doctor-Patient Relationship." The attendees are mavericks. Independent thinkers. There are over five hundred of them here, and there would be more, but the spaces sold out.

The gathering is being held by Front Line Covid-19 Critical Care Alliance (FLCCC), a growing group of physicians who joined together in April 2020 to try to cut through the censorship and false narratives launched against what many of them saw as effective treatments for their Covid patients. Treatments that their affiliated groups or the government prohibited them from giving.

As I've mentioned, FLCCC was started by Dr. Pierre Kory, a cutting-edge critical care researcher. He had a comfortable place in the mainstream until he refused to follow what he felt were irresponsible mandates on Covid treatments that restricted the type of care he was allowed to give his patients. It was much the same story for FLCCC cofounder Dr. Paul Marik. Before Covid, Dr. Marik was known for developing a lifesaving protocol for sepsis. He trained in a diverse set of specialties: internal medicine, critical care, neurocritical care, pharmacology, anesthesia, nutrition, and tropical medicine and hygiene. He taught medical school and led the Division of Pulmonary and Critical Care Medicine at Eastern Virginia Medical School (EVMS). In 2017, he received National Teacher of the Year award from the American College of Physicians. He's written more than five hundred journal articles. That now includes more than ten papers on treating Covid. Dr. Marik told me he left his hospital and regular practice after he was forbidden from giving his patients treatments he said worked for Covid, from vitamin C to ivermectin, a treatment that he's convinced saved lives. He says the next seven patients he saw in the ER without being able to treat them as he saw fit—died. He says he couldn't continue on in a system like that.

Though the FLCCC was created as an outgrowth of Covid controversies, it has expanded its areas of interest to cover the entire range of medical issues that establishment medicine has turned its back on. For example, instead of just medicating people who have diabetes, an immune disorder, these doctors want to get at the heart of why the disorder has exploded and how it can be prevented. Instead of putting Americans with now-common disease combos such as diabetes, high blood pressure, high cholesterol, and heart disease on four or more medications for the rest of their lives, the FLCCC doctors aim to look outside the prescription drug route to find preventive measures and cures. Instead of looking only to treat cancer and stand on the sidelines as its incidence explodes, this group seeks to truly identify and eliminate the multifaceted causes. This may mean funding studies that the drug industry won't sponsor because there's no pot of gold for it at the end of the rainbow. It certainly means the FLCCC has fallen under attack by the usual propagandists who do not want this kind of medicine to be practiced.

But there are signs that patients are demanding it. Frustrated by their doctors' responses, they are increasingly going outside their insurers and regular physicians' groups to find answers and get well—even if they have to find a way to pay out of pocket. It's leading to a quick expansion in the number of physicians practicing "root cause," "functional," or "integrative" medicine.

The FLCCC conference in Phoenix is a story of high viewer interest. In a nonmanaged news environment, hundreds of doctors gathering as part of a new movement to address our many health crises in novel ways would draw interest from many news organizations. The fact that I seem to be the only national TV news journalist to attend and report on it speaks volumes.

CHAPTER 16

New Beginnings

Carvey and Spade

When popular culture icons start lampooning the government's Covid missteps, you know a chord has been struck. At the end of June 2023, comic entertainers Dana Carvey and David Spade engaged in a satirically titled "We Miss Covid" segment in their *#FlyOnTheWall* podcast.

Carvey: I miss Covid.

Spade: I know. Dude, dude! You know when I knew there was trouble? When anyone that came to our country didn't have to get a vaccine. And I go, "If you're telling me I can't go to work, but everyone coming in doesn't have to get one?"

Carvey: Well . . . when Fauci said [imitates Fauci voice:], "Okay, I'm sorry, if you've had two boosters and two vaccines, you can get and give Covid to another guy who's had five vaccines and four boosters. . . . (It's just more vaccine but 'booster' sounds better.) Anyway, a guy with 25 vaccines would get and give Covid to another guy with 25 vaccines. That's why I'm introducing the daily Covid shot. Every day, you get a shot. By the time you get to your car, you got no immunity. But it's a beautiful 39 seconds!

Well, if you don't laugh, you'll cry. As you finish this book, I suggest there are two key takeaways.

First, the ethics lapses and fraud that happened during Covid—from the cover-up of its origins, to mismanagement of treatments, to the disaster of the vaccines—are on a scale larger than the world has ever seen before. We will be living with the aftereffects for the rest of our lives and, I'm afraid, so will our children and theirs. Just as with America's series of financial crises, those who created and profited from the systems that broke down seem to still be there running things. Not a single public health official has made so much as a mea culpa, given an apology, or even made a public acknowledgment that they were mistaken about the disease or vaccines. In other words: they'd do it all again. In the same way that the CDC made the farcical conclusion that its primary sin was just being *too* science-based, the official "lookbacks" of policy mistakes and misbehavior tend to deflect from the real issues. Congress had some hearings and left it at that. Nobody has been held responsible. No action taken. The need to dismantle and rebuild the CDC and, arguably, our entire public health structure, was conceded by even some of CDC's staunchest supporters. But those reforms didn't materialize. And the CDC, NIH, HHS, and FDA continue to receive collective billions of dollars in budget increases each year to help them accomplish more of the same.

The second takeaway is that all of this has prompted a call to action that's louder than most any before. More people are coming to understand that the perversion of science impacts virtually every aspect of our health. It's why chronic illnesses are running amok under the watch of the best-funded public health system on the planet. Why our government allows (and even mandates) that we eat food adulterated with cancer-causing chemicals, hormone disruptors, neurotoxins, and other poisons. We feed it to our children at home, and they're fed more of it at school. It's why our meat and fish contain harmful additives. Why our water is deemed safe to drink when it actually contains pesticides and other chemicals, and medicine that's leached in after being excreted into sewers from our overmedicated population. In this way, males are drinking water tainted with birth control medication, women are sipping on unnatural testosterone, and nearly all of us are inadvertently consuming some level of antidepressants, statins, diabetes medicine—whatever the population at large is ingesting. And none of this seems of concern to our government. *In what noncorrupt world does this make any sense?*

It's also why, when anomalies surface, such as the spike in autism or gender confusion, the cry to learn what could be causing them is muffled by industries seeking to normalize and deflect from the true sources.

Americans are increasingly depressed, fat, infertile, unfocused, and medicated. Among comparable developed nations, we're taking the most pills, paying the most for our medical care, and growing the sickest. Yet it's as if our health experts don't notice what's right in front of their noses. They remain hyperfocused on collaborating with the insurers, hospitals, and pharmaceutical interests that are getting rich off our poor health.

Doctors toil away in their stovepiped specialties, rely on propaganda passed off as medical studies and public health guidance, and shrug their shoulders when an illness baffles them. By design, they lack the time or curiosity to figure out what's really causing diseases like cancer or autism, and remain eager to pump out the latest pill or injection.

We're suckers. We continue to fund the broken system with record amounts of our tax money, while there never seems to be follow-through on serious oversight of the industries that give so generously to our media and politicians.

Fortunately, there are some tangible things you can start doing today to become well positioned to get at the facts.

There are steps you can take to expose the distorted medical system.

There are ways to fight back.

What follows is some practical guidance for truth-seekers looking for information rather than spin, and who are searching for honest scientific guidance when it can be elusive, ever-evolving, and always managed by unseen forces. I'm including specific sources for you to consider, as well as suggested actions that can help you become part of the solution.

Practical Guidance for Truth-Seekers

Assess the Track Record

Some rules that apply to divining truth in science are the same as those used to separate wheat from chaff in today's warped news landscape, as I've outlined at length in previous work.

One of the best strategies for success in each scenario is to identify who's proven accurate on key points in the recent past. For example, we now have the benefit of hindsight to see who was correct on fundamental controversies during Covid. I now give greater weight to those voices when they speak on emerging health issues, and I suggest you consider doing the same. Conversely, greater skepticism should be applied to the authorities who proved incorrect. While this may sound obvious, if you examine the post-Covid landscape, you'll see that many in media and medicine continue to illogically reference and quote the same authorities that gave the worst, most damaging advice. Big tech companies and fake fact-checkers in the media are still advancing the pharma-government line and suppressing dissenting information. Google and the other information manipulators are still finding billions of dollars to fund dishonest propaganda of untold proportions. Unreliable sources are relentlessly pushed on us from every direction. So, it's not always simple to sort through the noise.

Question the Prevailing Narrative

Along those lines, there are many sources you can consult specifically because the truth is frequently the opposite of what they claim. That can reveal a lot, too! When highly publicized "fact checks" and public health officials all seem to be saying the same thing, it's often a sign that the invisible hand of powerful interests is at work. Those interests seek to dismiss information they see as damaging and change the narrative. When so many in the media were disparaging hydroxychloroquine and ivermectin as Covid treatments, and smearing the doctors, researchers, and officials on that side, it meant to me that these drugs could actually have serious potential. It led me to dig into the question of: *Who wouldn't want people exploring the potential for these treatments and why?* During early Covid, when it seemed as if so many sick people were being put on ventilators and dying, our health officials were at a loss for what else to do, yet continued to press the same strategy. It meant to me that we probably ought to be doing something else. When the CDC changed overnight from saying that Covid isn't harmful to children, to insisting that all kids should be given experimental Covid shots; and when the media promoted this nonsense without so much as asking obvious, logical questions, it meant to me that something was up. These are commonsense thoughts that you may have had too. Trust them. It's easy to begin doubting yourself and your cognitive dissonance when you're surrounded by people who constantly insist that up is down. But their tactics only work if you buy into the fantasy.

Another way to identify questionable sources is by paying attention to the types of arguments they use. Do they advocate for censorship? If you're under age thirty, you may not realize that the drive to censor certain people and ideas en masse is a relatively new one here in America. Historically, censorship was seldom overtly promoted because it's contrary to the values under which the country was founded. Our forefathers specifically sought to eradicate tyranny over free speech, the kind of tyranny that had grown commonplace in England. Until not long ago, Americans were guided by the notion that bad ideas should be countered with better ideas—not censorship. More importantly, the government was constitutionally forbidden from having a role in defining the supposedly "bad" ideas and suppressing them. Otherwise, corrupt leaders would have the power to drown out dissent and oppress challengers.

But today, we know that American government leaders have interfered in our constitutional freedoms in the most invasive ways, then awarded themselves impunity. They've corrupted science, and crafted laws and public policies that enrich special interests. Then, the violators acquit themselves of their crimes.

In short, remember that the censors are never the good guys.

Today, when a political figure, fact-checker, or other information source seeks to censor, you should take it to mean the censored item *may be worth your consideration.* The same is true when an information source spends a lot of effort smearing people or groups rather than countering their ideas with different, well-argued ideas. And lastly, those who attempt to shut down discussion by claiming "the science is settled" are almost surely spouting propaganda. No real scientist uses the term "settled science" or shuts down discussion on scientific topics, where knowledge is constantly evolving.

Identify Propaganda Terms

If you take notice, you'll find that propagandists frequently reveal themselves through their use of hallmark terms. Look to see whether a scientific information source is using key buzzwords. We can assume the terms are chosen because some PR guru somewhere discovered they tested as most effective at swaying public opinion. Examples include claiming a scientific idea or researcher is "discredited," "bogus," "debunked," "quackery," "shoddy," "lies," "pseudoscience," "disgraced," "a conspiracy theory," or "a myth." Propagandists may refer to their targets as "nutty," "nutjobs," "baseless," "cranks," "kooks," "quacks," "anti-," "deniers," "hesitant," or "-phobic." They may claim "science" is "settled." Visit RespectfulInsolance.com for an exemplary case of a propaganda website that makes liberal use of these terms. Just for kicks, try conducting a Google search for the word "discredited" or "debunked." See which news sources use those terms a lot, and what they're discussing.

Be Skeptical of Narrative Dominance

Another way to identify false narratives and bad sources is by seeing what seems to be dominating in the media, on social media, and on the news. When you conduct an Internet search for health or scientific information, note how suspiciously homogeneous the results are. Be skeptical when there are dozens of "fact checks" of the same "facts," and when they use similar language and phrases, always reaching the same conclusions on matters of debate. Be wary when they try to tell you that something in dispute is not in dispute. Be aware when a person or idea they claim is "controversial" is only controversial because they made it so. The odds are zero that so many writers and journalists would independently draw the same conclusions and use similar language at the same time over and over again. So when you see these patterns, it suggests an orchestrated effort. It should lead you to wonder who would want you to think that way, and why. You should consider the possibility that valid scientific counterpoints or facts are being suppressed.

Broaden Your Consumption

Please note that I'm not suggesting you cut out sources of information. Quite the opposite. Broaden your dragnet rather than shrink it. It's a matter of how you use the sources. You'll actually find yourself best informed if you continue to absorb slanted information pushed out daily so that you can compare it against better sources. Staying versed on false narratives dispatched by the usual suspects will make it easier for you to identify trustworthy information.

The Best Medical Propagandists

Now that you're armed with some basic strategies for ferreting out good information, here are a few examples of propaganda sources that have proven reliably unreliable. Their sketchy track records teach us that when they "fact-check" or make a claim, it's worth considering the opposite of what they say as possibly true.

Wikipedia's health-related content is among the most frequently pushed and accessed in the world. Google consistently ranks Wikipedia pages high in search results and features Wikipedia's spin on medical topics. Since the popular "digital assistants" Siri and Alexa use Google, and Google refers to Wikipedia, Wikipedia is ubiquitous. Unfortunately, Wikipedia is one of the most prolific purveyors of propaganda and misinformation on matters of science and health. That's because those writing articles and doing the policing have conflicts of interest. For example, under Wikipedia's policies, biographies are supposed to be written "responsibly, cautiously, and in a dispassionate tone, avoiding both understatement and overstatement" and should "document in a non-partisan manner what reliable secondary sources have published about the subjects." But in practice, biographies of people who are off-the-pharmaceutical industry narratives on health issues tend to be viciously slanted in ways that violate Wikipedia's policies. Yet the behavior goes unpunished and uncorrected. When biographies of controversial health figures, such as the vaccine industry's Dr. Paul Offit, are filled with glowing accolades, and when they omit serious mention of controversies and missteps, it typically means paid interests or people with strong ideologies are controlling the page. Likewise, Wikipedia's anonymous agenda editors control pages on health topics that further one-sided and inaccurate information. Pharmaceutical representatives have gotten caught editing Wikipedia topics under pseudonyms to promote their company's medicine, "debunk" side effects, and attack competitors. When Wikipedia tells you something medical-related is "debunked," it may well mean the thing is actually true. When Wikipedia tells you somebody is "known for promoting" a "debunked" treatment or idea, take that to mean the treatment or idea may actually work. There are active propaganda efforts to make people think that Wikipedia's health information is unbiased and wholly reliable. But in practice, under the current model, that can never be the case.

Children's Hospital of Philadelphia/Merck's Dr. Paul Offit, and **Baylor College of Medicine Dr. Peter Hotez** have been some of the most-quoted and profuse propagandists speaking on the side of pharmaceutical interests. Even as they've frequently proven wrong or made false and libelous statements, the media continually returns to dip into the same well of disinformation. When these figures go on the warpath about a topic or against a person, the record teaches us that they may be covering for an uncomfortable truth.

A host of **blogs, bloggers, and websites** are on a daily mission to disseminate a singular narrative on health questions. Some of these sources use words like "science" and "skeptic" in their titles to try to convey a patina of credibility. You can count on them to sing from the same song sheet and smear those who dissent. They take the industry side in any discussion as if the non-industry views could never, under any circumstances, possibly have merit. They also claim to know the definitive truth on emerging health issues that nobody could have the final word on. Your instincts should tell you that whatever position they're taking, the opposite may be true. Here are some examples, though some of these propaganda sites have changed their names over time or moved on to the job of spinning on other health topics: Autism News Beat; Autism Science Foundation; HealthPartners.com; Left Brain Right Brain; Media Matters; Respectful Insolence, Retraction Watch, Science Based Medicine, and ScienceBlogs (David Gorski/"Orac"); The Science Post; Sense About Science; SethMnookin; Skepchick; Skeptical Raptor; The Vaccine Blog; Voices for Vaccines; The Vaccine Mom; Popsci; The Skeptic's Dictionary; The Skeptical OB; and Quackwatch.

Sadly, today, you can put most **news and quasi-news sources** in the category of those that should not be taken at face value on health matters. It doesn't mean that every article they publish is wrong or that all their reporters are bad. In fact, many organizations that rank among the worst offenders for health misinformation also have some very good reporters who work there. But those reporters are inevitably drowned out by their publication's indefensible editorial slant. Don't assume you are getting the truth, the whole truth, and nothing but the truth from these frequently biased sources: *The Atlantic,* CNN, *Daily Beast, Daily Kos, Forbes, Fortune, The Hill,* Huffington Post, Intelligencer, Mediaite, Michael Hiltzik of the *Los Angeles Times, Mother Jones,* MSNBC, *New York, New York Times, Politico,* Salon, Slate, Talking Points Memo, *USA Today,* Vaxopedia, Vox, or *Washington Post.*

Academic, journalism, and government bodies; professional

associations; nonprofits; "fact-checkers"; and "patient groups" have all been created or co-opted by propagandists on a massive scale, as we've discussed. For example, the American Cancer Society receives an undisclosed amount of funding from the very industries that make products that can cause cancer. Patient groups like "Vaccinate Your Family" are arguably fronts for pharmaceutical interests. The American Council on Science and Health (ACSH) furthers industry propaganda on nearly every health controversy you can name. So do universities, academic groups, and nonprofits that receive funding from special interests to conduct "fact-checks" or produce "media literacy efforts" and create "media resources" that always take the side of the pharmaceutical industry and frequently distribute misinformation. Some of the offenders include the FDA, the CDC, the World Health Organization, Annenberg Public Policy Center, American Heart Association, Credibility Coalition, American Academy of Pediatrics, American Medical Association, Center for Countering Digital Hate, Columbia University, Duke Reporters' Lab, Facebook Journalism Project, Facebook's Lead Stories, FactCheck.org, Google and Google News Initiative, Health Feedback/Science Feedback, Health Policy Watch, Institute for Strategic Dialogue, International Fact Checking Network, Knight Foundation, Medscape, PolitiFact, Poynter, Snopes, the University of Pennsylvania and many other academic institutions, WebMD, Verify or VerifyThis.com, and YouTube, to name but a few. Many of these groups claim to be unbiased, neutral, science-based, or fair—but remember, there's no law that requires propagandists to describe themselves honestly and accurately. Just because they say they're something doesn't make it true.

Take the example of "Health Policy Watch," which touts itself as "a non-profit, open-access journal" that provides "Independent Global Health Reporting." It operates with support from the pharmaceutical-founded Wellcome Trust and a range of other left-leaning foundations, philanthropies, professional federations, and academic institutions. Its editor in chief, as of this writing, is Elaine Ruth Fletcher, a climate change activist who came from the left-leaning World Health Organization. The group's number two, Kerry Cullinan, came from the left-leaning "Open Democracy." Another group, the Institute for Strategic Dialogue, claims to be "a fiercely independent global organisation dedicated to powering solutions to extremism, hate and disinformation." However, it's driven by and partners with many of the same names that pull strings at so many other groups. These include Google, Facebook, YouTube, Amazon, Spotify, Microsoft, Bill & Melinda Gates Foundation, United Nations, Department of Homeland Security, and US State Department.

Recommended Sources

Next, some suggested sources for you to consider in order to learn what the propagandists are trying to hide. It's no accident that if you google most of the sources in this section, you will be told they are "controversial," "discredited," or "debunked." Most of them aren't controversial for any reason other than that the propagandists have sought to make them so in hopes that you'll disregard them. Ask yourself why the CDC and Dr. Fauci, who are genuinely controversial because of their conflicts of interest and misinformation, are rarely, if ever, referred to as "the controversial CDC" or "the controversial doctor Anthony Fauci." On the other hand, sources with sound records and perfectly reasonable scientific positions that run against the pharma grain are frequently called "controversial."

Keep in mind that you probably aren't going to find many infallible sources. The uncertain and evolving nature of science and medicine makes perfection nearly impossible. Instead, what you're looking for are good sources that acknowledge uncertainty and, when proven wrong, admit and self-correct.

One bold example of that is a medical student and researcher named Kevin Bass. In statements published in January 2023, Bass confessed, "I staunchly supported the efforts of the public health authorities when it came to COVID-19. I believed that the authorities responded to the largest public health crisis of our lives with compassion, diligence, and scientific expertise. I was with them when they called for lockdowns, vaccines, and boosters. I was wrong. We in the scientific community were wrong. And it cost lives. . . . It doesn't matter much, but I wanted to apologize for being wrong." Why aren't those in more established and esteemed positions willing to admit their errors? If there's something more dangerous than being so wrong on life-and-death issues, it's sticking by your mistakes at any cost.

If you want to get the industry's take on any of the sources I'm about to suggest, just google their names or search on Wikipedia. As I mentioned, you'll find a suspiciously uniform series of "fact checks" and "news" articles discrediting them. The list of recommended sources that follows is far from exhaustive. It's a starting point.

Dr. John Abramson is a family physician who has taught health-care policy at Harvard Medical School for more than a decade and a half. He proved to be a fair and wise voice that was front and center on the dangers of the painkiller Vioxx and cholesterol-lowering statins when many doctors were pushing them on a massive scale. He has testified as an expert in litigation involving pharmaceutical fraud. Now, when Dr. Abramson weighs in on a medical issue, I pay attention. He is author of *Sickening: How Big Pharma Broke American Health Care and How We Can Repair It* and *Overdo$ed America.*

The Association of American Physicians and Surgeons (AAPS) is a nonpartisan professional association of physicians that serves as an alternative to the left-leaning American Medical Association. Often smeared by pharmaceutical interests as "conservative" and promoting "quackery," it has proven reliable on topics riddled by misinformation elsewhere. You can find both a newsletter and a journal at the website aapsonline.org.

Dr. Jay Bhattacharya is a professor of health policy and an infectious disease epidemiologist at Stanford Medicine. He coauthored the Great Barrington Declaration, taking an early stand against the unscientific Covid lockdowns, rightly pointing out that such an approach had long been recognized as likely to do more harm than good. Nearly a million people signed on to the Declaration, including tens of thousands of medical professionals. However, the Declaration and its supporters were subjected to orchestrated smears at the hands of

Drs. Fauci and Collins at the National Institutes of Health. Dr. Bhattacharya's coauthors on the Declaration are **Dr. Sunetra Gupta** and **Dr. Martin Kulldorff.** Dr. Bhattacharya publishes under a Substack entitled "The Illusion of Consensus" at substack.com with journalist **Rav Arora.**

Brownstone Institute is a freedom-based nonprofit founded in May 2021 in response to the government's actions during the Covid pandemic. It describes its vision as one of "elevated learning, science, progress, and universal rights to the forefront of public life." It publishes an eclectic mix of writers and writing at brownstone.org.

C-SPAN is a valuable resource for unfiltered information of all kinds. You can find a bountiful archive of interviews, events, and press conferences uninterrupted by the normal spin. You can compare what you see there to how media outlets report the same event, and begin to understand the media's slant on a given topic. When I'm looking for a specific archived event or topic, instead of going to the C-SPAN page (c-span.org) and using the search function there, I find I'm more likely to locate what I'm looking for when I do a general Internet search using the term "CSPAN," the topic, and if known, the date.

Tucker Carlson is a former Fox News host who now conducts insightful interviews that include health scandals like we've discussed in this book. His interviews appear on "X," formerly known as Twitter, and are viewed by millions under his account @TuckerCarlson.

Children's Health Defense is a nonprofit group founded by Robert F. Kennedy Jr. It draws attention to child safety issues, including vaccine dangers. The group has a searchable database of hundreds of peer-reviewed, published articles on environmental contaminants implicated in modern childhood diseases and epidemics. You can access the material by going to childrenshealthdefense.org and clicking on the "science" tab.

Cochrane Collaboration describes itself as "a global independent network of researchers, professionals, patients, carers and people interested in health." While it's had its share of controversies, and firmly marches to the drumbeat of the medical establishment on some topics, it has also been known to provide valuable, off-narrative scientific information. Go to cochrane.org, click "Cochrane Library," and scroll down to topics of interest.

Epoch Times is an independent, subscription-based news source that has conducted a steady stream of impressive investigative reporting on health-related issues. Like other off-narrative sources, propagandists have falsely labeled it as a "conservative" publication, and it has been the target of orchestrated smears by the interests it exposes. You can subscribe at epochtimes.com.

Front Line Covid-19 Critical Care Alliance (FLCCC) is a nonprofit founded by **Drs. Pierre Kory** and **Paul Marik**. The website provides patient resources, a list of doctors (under "providers"), and medical evidence (under "studies") that reference recommended Covid and Covid vaccine injury treatment protocols. Find the site by searching "FLCCC" or visiting covid19criticalcare.com.

Full Measure with Sharyl Attkisson and **SharylAttkisson.com** are two places where I publish my off-narrative reporting, including on health issues. At SharylAttkisson.com, you can click the "Health" tab for a list of reporting and resources on everything from "Vaccines and Medical Links" to Covid-related material. Also at SharylAttkisson.com, you can click the "Full Measure" tab and access an archive of all my TV program's cover stories organized by topic. One of the headings there is "Health-related." At the same "Full Measure" tab, you can access a list showing the TV stations that air *Full Measure* each Sunday, and see the program times in various cities. *Full Measure* feeds to 43 million US households every week. To watch replays of the program anytime, go to FullMeasure.news online.

Glenn Greenwald is a Pulitzer Prize–winning investigative journalist and attorney who broke the Edward Snowden story about the government's massive, secret spy operation against American citizens. He later cofounded a journalism group, The Intercept, then quit when his own organization censored his reporting on Biden family scandals. He's been a bold voice questioning the government narrative on health matters and other controversies. He has a live online news report on Rumble weeknights at 7 p.m. EST at rumble .com/GGreenwald.

Jeremy Hammond is an independent journalist who was heavily targeted and censored for his critical reporting on US government policy on Syria, and later on Covid lockdowns, natural immunity, and more. On his website, Hammond writes, "I expose dangerous state propaganda serving to manufacture consent for criminal government policies." He has a free newsletter and a free e-book entitled *The New York Times vs. Robert F. Kennedy, Jr.: How the Mainstream Media Spread Vaccine Misinformation!* His articles are available at JeremyRHammond.com.

InfluenceWatch.org and **Influence Watch podcast** are products under **Capital Research Center (CRC)**, which examines money influences in charities, politics, and media from a free market perspective. Influence Watch is intended to provide "more fact-based, accurate descriptions of all of the various influencers of public policy issues." According to the group, "We let the information speak for itself—information that frequently is not cited in reports about these individuals and organizations." If you're wondering about a widely quoted charity or group that's glowingly described in Wikipedia or others in media, it's worth searching for the name of the group at Influence Watch to find any possible hidden information. You might be surprised by what you find.

Informed Consent Action Network (ICAN) was founded by journalist and film producer **Del Bigtree**, host of *The HighWire with Del Bigtree* Internet news show and podcast. Bigtree is known in part for walking away from his producing job at the CBS television show *The Doctors*, for which he received an Emmy Award. He says he left the program because of its pro-pharma tilt. ICAN has had success suing the CDC, the FDA, and NIH over issues of health and informed consent. To find out more, visit icandecide.org.

The Institute for Pure and Applied Knowledge (IPAK) research institute and IPAK-EDU LLC were founded by research scientist **James Lyons-Weiler**, PhD. They are designed to provide impartial research results and viewpoints on some of the most important and controversial topics in biomedicine, psychiatry, and sociology. According to the website, "Conducting science in a way that is independent of any profit motive—so the results can be better trusted—will be the core paradigm of the institute." IPAK features "Research conducted by people who, under the by-laws, can have no financial stake in the outcome." IPAK-EDU offers in-depth courses from the fundamentals to advanced topics in biology analytics, health, mind science, and the humanities. Visit ipaknowledge.org or visit its educational course site at ipak-edu.org.

Dr. Martin Makary is a surgeon and public policy researcher at Johns Hopkins University. He has written about why so many public health officials avoided talking about the power of natural immunity during Covid. He has also spoken to what's happened to science in the current manipulated information environment. He recommends Covid vaccines for many but has expressed opposition to vaccine mandates. A list of his books can be found at martymd.com.

Dr. Aseem Malhotra is a world-renowned British cardiologist who first drew the ire of government and industry forces when he put the focus on the role our adulterated diet plays in chronic diseases. He was an early adopter of Covid vaccines, but says when he watched his own father die from the vaccine, he did an about-face. On other health topics, he recommends people make dietary changes like limiting sugar and ultraprocessed foods to improve their health and reduce risk of heart attack and stroke. He also campaigns against overprescribing of medicine. More about his current work can be found at doctoraseem.com.

Dr. Peter McCullough is another esteemed cardiologist who boldly spoke out about Covid vaccine heart risks and advocated for medical freedom. Rather than be silenced, he gave up his job at Baylor University Medical Center, where he was vice chief of internal medicine. He has a Substack entitled "Courageous Discourse" and a foundation dedicated to "Advancement of clinical science, protection of personal autonomy, liberty, and constitutional rights." He's coauthor of the book *The Courage to Face COVID-19: Preventing Hospitalization and Death While Battling the Bio-Pharmaceutical Complex*. When he is interviewed or speaks to a medical issue, his views are well worth considering.

National Vaccine Information Center (NVIC) is a well-established information resource "dedicated to preventing vaccine injuries and deaths through public education and securing informed consent protections in public health policies and laws." Its website provides valuable information on vaccine benefits and risks, and includes otherwise difficult-to-find information such as specific vaccine lot numbers that have proven problematic or been linked to clusters of adverse events. You can visit at nvic.org.

Open the Books is a deep-dive research group that tracks government spending. It has a wealth of oversight reports, including ones about conflicts of interest in medical spending. It also has an impressive, searchable database of federal and local spending of taxpayer money. The group is built on the idea that "it's your money, and you deserve to see where every dime is spent!" Go to openthebooks.com and click "Reports" or use the search bar for topics of interest.

Chris Plante is a journalist-turned-talk-show-host and a former colleague from my CNN days ("when CNN was a news organization," as we like to say). There's no smarter or more wry analyst of political developments, which often veer into health and medical scandals. Plante has a daily radio show originating on WMAL in the Washington, DC, area, and a prime-time TV program on Newsmax.

Podcasts and Internet shows: There are countless podcasts and Internet programs that address health information in an informative and responsible way, airing a variety of viewpoints that are often censored in other forums. Two popular podcasts that have proven enlightening on these topics are ***The Joe Rogan Experience*** and ***The Adam Carolla Show***. A popular off-narrative Rumble program is **Russell Brand**'s live weekday show. The hosts are entertainers, not journalists or medical experts, and that may be why they're so willing to stray from popular narratives—in a good way.

Project Censored is a *real* media watchdog, media literacy group, and independent journalism project—not a fake one like so many. Under the leadership of **Mickey Huff**, this left-leaning nonprofit is one of the most solid challengers of the pharmaceutical establishment, government conflicts of interest, and false narratives in media.

Whether it's Project Censored's radio show, its annual list of "most censored" news stories, or its films, there's a lot to learn from the group. To connect, visit ProjectCensored.org.

Public Citizen is a left-leaning, politically active watchdog group originally founded by consumer advocate Ralph Nader. But what deserves attention here is Public Citizen's "Health Research Group." It's often on the leading edge of drug safety issues and scandals, and is active in petitioning government agencies and Congress on medical controversies. When Public Citizen's health experts weigh in on a debate, I find their views are always worth hearing. Go to citizen.org and click "health care" under the top left menu icon or go directly to citizen.org/topic/health-care. You'll find all kinds of reports, letters, and information.

RealClearPolitics is a valuable resource on many topics. Its investigative arm conducts hard-hitting, well-documented investigations that may be fearlessly off the establishment news narrative. What you might find of particular value is its "Fact Check Review" and searchable database. Unlike fake "fact check" efforts we've discussed, the project at RealClearPolitics really gets to the heart of facts without the spin. According to the website, "The goal of the Fact Check Review project is to understand how the flagship fact-checking organizations operate. What types of claims do fact-checkers review? How often do they fact-check Democrats or Republicans? What claims are labeled 'misleading'?" **Kalev Leetaru**, a senior fellow at the George Washington University Center for Cyber & Homeland Security, is a key writer at *RealClearPolitics* devoted to "checking the fact checkers." Find the searchable database at realclearpolitics.com/fact_check_review or simply search "Real Clear Politics" and "Fact Check Review" online.

Dr. Harvey Risch is a Yale epidemiologist who wrote and spoke about the "loss of civil liberties during Covid" and what he called the government's "intentional mismanagement" of the pandemic, suppression of early treatment, and needless lives lost. He also dissected the propaganda campaign against hydroxychloroquine. He publishes on **America Out Loud** at americaoutloud.news and Brownstone.org.

Aaron Siri is an attorney and managing partner of Siri & Glimstad.

He has filed important Freedom of Information Act requests and mounted legal challenges against "mandated medicine." He has a Substack called **Injecting Freedom** devoted to "[s]afeguarding individual rights demands constant legal, social, and political struggle against government censorship, coercion, and mandates."

John Solomon is an investigative reporter who has fearlessly broken a great deal of news on a variety of topics, including health-related scandals. You can find his work and podcast by visiting his online newsgroup: **JusttheNews.com**.

Substack is a rapidly growing alternative information platform that allows many voices to be heard without the trademark censorship that's become so commonplace elsewhere online. Here, you can curate your own sources and find many journalists and scientists publishing engaging opinions, information, and science.

Paul Thacker writes one of my favorite Substacks newsletters: **The Disinformation Chronicle**. He's a journalist and former investigator in the US Senate, where he exposed corruption and conflicts of interest in science and medicine. Thacker has written for the *New York Times, BMJ Investigations, JAMA, Washington Post, NEJM,* the *Los Angeles Times, The New Republic,* Slate, and *Mother Jones*—and survived, despite being off the narrative on Covid and other important health topics. His exposé about a Pfizer whistleblower on the Covid vaccine trials was published in the *British Medical Journal* and received first place in the inaugural Sharyl Attkisson ION Awards for investigative and off-narrative journalism in 2022.

US Right to Know is a nonprofit public health research group that has published a great deal of valuable information on health conflicts of interest and the dangers of chemicals and ultraprocessed foods. It has also investigated the origins of Covid-19, risks of gain-of-function research, and mishaps at biolabs. There's a wealth of information on a wide variety of science topics at usrtk.org.

Vaccine Court is the little-known, specially created court that's part of the US Court of Federal Claims in Washington, DC. It's where vaccine-injured patients are required to file any court claims they wish to make. You can read some of the cases by going to govinfo.gov. In the search bar, type "Court of Federal Claims" along with terms such as "vaccine," or a specific vaccine such as "influenza vaccine." You can also search the same way using terms such as "autism," "encephalitis," and/or the name of a victim. Visit the vaccine compensation page at hrsa.gov/vaccine-compensation and under "Resources" you can explore how much money has been paid out for what type of vaccine injuries.

Action Plan

The government and industry have been so successful in plastering our lives with their narratives and bullying those who dissent that it can be easy to throw up your hands and feel you can't make a difference. But remember that your complacency, silence, and discouragement are the means to their success. When you care, speak out, and become motivated, it dooms their efforts to failure.

Now that you're equipped with some good information sources, you can become an advocate for open discussion of the topics the propagandists seek to slant. **Participate in dialogues** to let others know that when they have commonsense questions about the supposed "truth," they're not alone. **Raise questions** in civil conversations. **Call out the tactics of propagandists.**

Another action you can take is to **challenge studies** that are slanted or wrong. Raise questions or write about the errors on social media. Contact editors of the operative medical journals and pursue corrections or retractions. You can bet propagandists are paying big money to have their experts poke holes in and demand retractions of studies that harm their interests. But too many bad studies remain published and uncorrected because nobody effectively challenges them.

With some effort, it is possible to expose unscientific propaganda and even force retractions. An article published on December 22, 2022, that claimed Donald Trump was "the Main Driver of Vaccine Misinformation on Twitter" was eventually retracted as misinformation. An extraordinarily lengthy explanation raised concerns about "the methodology, results, and conclusions."

Serious ethical lapses in the Baby Oxygen study discussed near the start of this book were originally exposed not by layers of official reviewers but by a layperson. In September 2022, a taxpayer-supported study at Temple University with Janssen Pharmaceutical was retracted after an ordinary reader raised questions. The study concluded that the widely used blood thinner Xarelto could have a healing effect on the heart. But there was alleged manipulation of important images and data. Two of the fourteen authors were employed by Janssen, and one of them owned Johnson & Johnson stock. A third author received "personal fees" from Janssen. All of the involved scientists who have spoken have denied wrongdoing. But as we've discussed, it's exceedingly common for academic studies held out as independent to be what amounts to products of drug companies.

You can **file Freedom of Information Act requests** to obtain documents regarding taxpayer-funded studies or other material held by public agencies. Be persistent: they rarely make it easy for us to see the information we own and frequently don't follow Freedom of Information Act law. Contact representatives in Congress to tell them what you find, explain what you care about, and demand action.

And finally, an important action you can take is to **add to the greater body of knowledge by properly reporting any illnesses** you get after a vaccine or another medicine, and encouraging others to do so. I'll give details on how to do that in a moment. First, let me explain why it's important.

If you take medicine, your doctor should regularly ask if you've developed any new symptoms or adverse events. If your answer is "yes," then the doctor should file a report with the appropriate federal database. But most of them don't ask the question or report the results. Many erroneously think that in order for an illness to merit reporting, they're supposed to "know" it was "caused" by the medicine. That's untrue—and would defeat the purpose of the reporting system, which is designed to capture *all* events in order to see if there are patterns that imply a relationship and are worthy of investigation.

The fact that medical professionals misunderstand or misrepresent how the adverse event reporting system works is yet another result of biased medical school teachings. It exposes yet another failing of our government and medical establishment. A system truly interested in improving public health would make reporting and analysis of this data a top priority.

During the Covid pandemic, I interviewed a physician assistant named Deborah Conrad, who said her hospital started treating a sudden flood of very sick people—a lot of them young—for apparent Covid vaccine-adverse events. Conrad began reporting them, as required, to the federal database. Pretty soon, she told me, nurses and doctors who didn't have time to file their own reports on patients, or didn't know how, brought their cases to her and asked her to file the proper reports with the federal database. She says there got to be so many, she could barely keep up. Before long, she says, hospital administrators called her in for a talk and told her to cut it out. She says she reminded them they were bound by professional rules to file the adverse event reports. They had no choice. But when she persisted, she was fired. Public health was truly turned on its head during Covid. Health professionals following rules designed to put patient safety first were treated like criminals.

The federal adverse event reporting system is paramount because it's the method by which most dangerous side effects are ultimately detected. They aren't unearthed in the short-term studies used to approve a medicine. They're captured through the post-marketing reporting system that casts a broad net. Who would have originally guessed that Viagra causes blindness? Unfortunately, most doctors prescribing that erectile dysfunction drug were waiting for someone to tell them that it impacts vision rather than using their own patients' experiences to discover that fact on their own.

In order for the adverse event reporting system to work, it's key to report not only illnesses that occur within a few days or weeks of someone starting a medicine or getting a vaccine, but also those that crop up months or years later. One study finds that life-threatening problems with drugs can begin to surface, on average, about four years after their approval.

But how many times has your doctor quizzed you to learn if you developed any new symptoms after taking a vaccine or medicine? How many times when you've ended up sick at a doctor's office or at a hospital has a medical professional taken the time to explore whether there could be a connection to medicines you take? Probably not often.

That explains why experts say the federal database contains only a small fraction of the actual adverse events a medicine may cause: most doctors fail to properly report. Therefore, scientists say, each single adverse event that does get reported represents 1,000 to 100,000 times more that go unreported. This also explains why a mere handful of deaths that are officially connected to a drug are enough to sound alarm bells. Because 12 known serious injuries imply there could really be 12,000 to 1,200,000.

So now that you know how important it is to make sure you're counted, here's how to do that.

MedWatch is the name of the FDA system under which consumers can report injuries and/or deaths that occur after nonvaccine medicines or use of medical products. Unfortunately, the way MedWatch is set up is very confusing, and the reporting form is difficult to find. If the government really wanted to collect as much data as possible, it would devise a system that's a lot easier for us to navigate. Anyway, an Internet search for the MedWatch online reporting form will likely lead you through an arcane process that prompts you to download a PDF file to fill out. Not much help.

Here's the easier way to file online. Even this is far more complicated than it should be, but after you do it once, it becomes pretty quick and easy.

First, search the Internet for the phrase "MedWatch Forms for FDA Safety Reporting." Click on that page and scroll down to the option titled: "MedWatch Online Reporting for Health Professionals, Patients and Consumers." (It's the fourth line down, as of this writing.) There, click the drop-down arrow to see "Online Voluntary Reporting." Click that, and then go to the "Begin Online Report" box. Click "Consumer/Patient" and you're there. Or you can go directly to the form via this link: https://www.accessdata.fda.gov/scripts/medwatch/index.cfm?action=consumer.reporting1.

Vaccine Adverse Event Reporting System (VAERS) is the federal database designed to gather reports of illnesses after vaccination. Historically, the government works with vaccine interests to cover up and explain away emerging patterns rather than address them in a constructive way or issue alerts to consumers.

There are some important facts for you to know about VAERS. Anyone can report an adverse event; it doesn't have to be the patient or doctor. You do not need all the information the form asks for, such as the brand name of vaccine or specific lot number; just fill out the parts you know. You should file a report even if you think a doctor may have already done so; the system will cross-reference and correct for duplicates, as necessary. A VAERS report should be filed after every illness that occurs after vaccination regardless of whether a doctor thinks the vaccine caused the illness. Neither the patient nor physician is supposed to try to determine or exclude the vaccine's role. They don't have the information or expertise to do that. That's what the database is for. Lastly, don't forget that scientists have documented that adverse events from vaccines can occur months or years after the vaccination. Go to vaers.hhs.gov and click the "Report an Adverse Event" tab to fill out a form online.

Don't worry that you may be reporting a symptom that turns out to be unrelated. The federal databases were designed with the understanding that many reports will not, in the end, be due to a medicine. But only when a lot of illnesses are recorded can unusual or suspicious trends emerge for further investigation.

The last action I'll suggest you can take is to **increase awareness of federal Vaccine Court**. Yes, it is very tough to get compensated for a vaccine injury. But with proper contemporaneous documentation, it can be done. You can find a list of lawyers who understand how this complex and unusual court works by searching "United States Court of Federal Claims, Vaccine Attorneys" online. As vaccine cases work their way through court—win or lose—they can add to the body of knowledge on the consequential subject of vaccine injuries.

Epilogue

Would you believe it if I told you that in this book, we've touched upon only the tip of the proverbial iceberg? I've tried to boil down my own journey toward enlightenment in a succinct way that represents the long road, but with enough detail to get a sense of the thoughtful and time-consuming process it has been. I've guided you down a rabbit hole so deep that it's taken us through many enlightening, interconnected chambers. And now, you're part of a new alliance: a true, grassroots coalition.

A parent and a brother of mine earned medical degrees. Like most of you, I grew up unquestioning of doctors.

As a patient, I didn't believe the government or medical professionals would ever lie. I thought that anytime I went to the doctor, I would get the very best help in order to get better.

As a parent, I believed that doctors, government medical experts, and child safety groups put my child's health first, above all other concerns. I made sure my child got all the recommended vaccines without considering that they might not be necessary, might not work, or could do harm.

As a young reporter, I believed pretty much whatever message the medical establishment advanced. I thought scientific research was pure and to be taken unquestioningly at face value. Without scratching the surface, I bought into the propaganda telling me that a lot of medical concerns amounted to myths or conspiracy theories.

A lot of journalists allow their preconceived notions and uneducated or uninformed perceptions to drive their reporting. That's a big flaw in our profession. Journalism schools don't do a good job of teaching students how to be receptive to information that may go against what they personally believe. The fact is, when we let go of our own ideas and open our minds, we can learn a lot more and get a lot closer to the truth. Rather than becoming vested in using our stories to prove our own points, we should be willing to listen to and present views that differ from what we think we know. We must also learn how to cut through the powerful propaganda machine that inserts itself into nearly every aspect of our lives. Letting go of our own ideas when warranted, or admitting they were wrong, can be among the most rewarding intellectual exercises and journalistic endeavors we can undertake.

When it comes to health reporters, and beat reporters in general, they tend to develop a form of Stockholm syndrome. (Stockholm syndrome refers to hostages developing a bond with their captors.) Beat reporters admire or become friends with the officials they're covering. Or they form a mutually beneficial relationship whereby each one succeeds more if the reporter buys into the spoon-fed narratives and press releases without considering ulterior motives and alternate views. At CBS News, one of my best bosses assigned me to cover medical controversies, telling me he was purposely letting me ruffle feathers at the CDC, the FDA, and NIH, while letting the assigned medical reporter remain on good terms with the agencies so she would be granted interviews and have her calls returned. Classic good cop, bad cop.

I personally found that handouts from the federal agencies are generally of little to no news value. They're giving you information when they want you to have it and the way they want you to see it. Partial truths. Sometimes: outright lies. It never made sense to me that news organizations seem to believe their beat reporters must stay on the best of terms with those they're covering in order to get good information. In reality, all they're getting is spin. Maybe if they're "in good" with the agency, they'll be first to get the spin. That's nothing to be proud of. I learned I was no less likely to get information of value, and was sometimes more likely to get it, when the agencies weren't happy with me or my reporting. When they knew I wasn't going to unquestioningly report what they wanted me to report.

All of this is probably why I ended up on a journey that would transform my professional life, bringing clarity to some of the most fascinating and perplexing questions in public health. As an investigative journalist I've learned how money and profits drive virtually every decision impacting public health. I've discovered the hard way that government officials, doctors, medical associations, researchers, and nonprofits mislead and lie. I've come to understand that the many scandals occurring in the health arena involve the same components as the other scandals I've investigated over the decades: cover-ups, greed, profits, and lies.

As you've come to discover your own truths about public health officials, doctors, and the science they produce or rely on, it can be extremely unsettling. But you're actually on a much better path. You're taking control of your health and that of your family.

For others, there may be no way to awaken them. They may find comfort in the delusion that the government is here to help, our medical professionals are infallibly altruistic, public health experts would never hurt anyone, and science is incorruptible.

But rest assured as you're reading this, millions of people have begun to open their eyes. Millions more are learning. The propagandists have become so audacious, the censors so overreaching, that their tactics are backfiring in real time.

Thank you for deciding not to look the other way. Thank you for choosing the more difficult but rewarding path.

Now let's get busy.

Index